William R. Park
CONSTRUCTION BIDDING FOR PROFIT

J. Stewart Stein
CONSTRUCTION GLOSSARY: AN ENCYCLOPEDIC REFERENCE
AND MANUAL

James E. Clyde
CONSTRUCTION INSPECTION: A FIELD GUIDE TO PRACTICE

Harold J. Rosen and Philip M. Bennett
CONSTRUCTION MATERIALS EVALUATION AND SELECTION:
A SYSTEMATIC APPROACH

C. R. Tumblin
CONSTRUCTION COST ESTIMATES

CONSTRUCTION COST ESTIMATES

Girder pour
Bakersfield Memorial Stadium
Contractor: Tumblin Company

CONSTRUCTION COST ESTIMATES

C. R. TUMBLIN, P. E.

A Wiley-Interscience Publication

JOHN WILEY & SONS

New York • Chichester • Brisbane • Toronto

Library of Congress Cataloging in Publication Data

Tumblin, CR   1917–1979
  Construction cost estimates.

  (Wiley series of practical construction guides)
  "A Wiley-Interscience publication."
  Includes index.
  1. Building—Estimates. I. Title.
TH435.T85   692'.5   79-16376
ISBN 0-471-05699-5

Printed in the United States of America

10 9 8 7 6 5 4 3 2 1

Series Preface

The Wiley Series of Practical Construction Guides provides the working constructor with up-to-date information that can help to increase the job profit margin. These guidebooks, which are scaled mainly for practice, but include the necessary theory and design, should aid a construction contractor in approaching work problems with more knowledgeable confidence. The guides should be useful also to engineers, architects, planners, specification writers, project managers, superintendents, materials and equipment manufacturers, and, the source of all these callings, instructors and their students.

Construction in the United States alone will reach $250 billion a year in the early 1980s. In all nations, the business of building will continue to grow at a phenomenal rate, because the population proliferation demands new living, working, and recreational facilities. This construction will have to be more substantial, thus demanding a more professional performance from the contractor. Before science and technology had seriously affected the ideas, job plans, financing, and erection of structures, most contractors developed their know-how by field trial-and-error. Wheels, small and large, were constantly being reinvented in all sectors, because there was no interchange of knowledge. The current complexity of construction, even in more rural areas, has revealed a clear need for more proficient, professional methods and tools in both practice and learning.

Because construction is highly competitive, some practical technology is necessarily proprietary. But most practical day-to-day problems are common to the whole construction industry. These are the subjects for the Wiley Practical Construction Guide.

M. D. Morris, P.E.

New York, New York

v

Preface

This book presents a practical method for preparing consistent and detailed cost estimates for construction projects. The writer is a general contractor with more than 35 years of successful business experience bidding and constructing engineering and building projects. The procedures illustrated are the ones used, with minor variations, throughout the industry.

Although the material is written primarily for upper division students in engineering or architecture who plan to enter the construction industry, it is also intended as a reference book for general contractors and for those professionals who are involved with the preparation of job plans and specifications and with the awarding of construction contracts.

Every construction project consists of a number of separate categories of work called job items. The estimating process requires each job item to be analyzed, divided, and subdivided, as necessary, to arrive at those primary activities that must be performed to complete the various operations involved.

The quantities of the primary activities and materials required are collected and summarized on the take-off for the item. The estimated cost for the item is then determined by applying a system of unit costs to these quantities. The unit costs for labor and equipment are selected from an array of unit costs or production rates achieved on past jobs involving operations similar to the ones being considered.

Each contractor's unit costs for labor and equipment are unique, and reflect his efficiency and experience, the job conditions relating to his cost data, the amount of job supervision used, and the quality of his cost-keeping system.

The recognition and the quantification of the required primary activities is of the utmost importance, since they must be priced to estimate job costs, timed to estimate the time required for completion, and monitored to control the cost and the progress of work under construction.

In the following chapters, typical job items are analyzed and the primary activities relating to the various types of work are identified. Take-offs are made illustrating the methods used to collect and summarize the quantities involved. Unit costs are presented, and these unit costs are indexed for conversion to those wage and rental rates applicable to the job being bid. Cost estimates as well as estimates of the time required for completion are illustrated for various job items and for assumed projects consisting of various combinations of job items. The estimates are identical to those made in practice; the job cost estimates include the costs of direct job overhead, general overhead, payroll expenses, and profit.

The unit costs presented herein cover those job items normally performed by general contractors, and are quite inclusive. The role of the subcontractor is discussed in detail, but unit costs for subcontract work are not covered. In the job cost estimates illustrated, assumed quotations from subcontractors are used in conjunction with the cost estimates made for those items of work normally performed by the general contractor.

The material is presented so that by the time you have progressed through part of the material covered in Chapter 4, you will be able to prepare "dry run" bids for small construction projects, and these bids can be compared with the actual bids submitted by the competing contractors. Such an exercise will demonstrate your proficiency in applying the methods illustrated, but your estimate cannot be considered reliable or sound, since field and office experience are required to properly modify average unit costs to suit particular job conditions. In addition, such exercises will not furnish any indication of your ability to actually do the work at the estimated cost, and you will not experience the tension and excitement that prevail on bid day, when the bid is backed with money and submitted against an absolute time deadline.

The ability to analyze the work going on about them will make the early years of on-the-job apprenticeship much more productive for those entering the construction industry, both in terms of their own professional development and in terms of their value to, and rate of advancement by, their employers. It will also enable them to start a practical file of job unit costs based on their own job experiences. A file of such unit costs is a basic necessity for all construction cost estimators.

Those who enter the construction industry will discover a world of unlimited opportunity. With each job well done, they will experience that quiet pride and inner satisfaction known only to those who build enduring things; and when their days in construction are over, their works will stand in silent tribute to their early struggles, to their efforts, to their abilities, and to their creativity.

C. R. TUMBLIN

Bakersfield, California
July 1979

Contents

2 Unit Costs

10 Base Courses and Pavements 352

11 Clearing and Grubbing—Finishing Roadway 367

12 Drain Pipe 371

Tables

Abbreviations

A	Area
BCY	Bank cubic yards
BF	Base frame or base frames; backfill
BF & E	Base frame or frames and extension frame. Thus 2 BF & E refers to a tower 2 base frames plus an extension frame in height
C	Carpenter
CF	Cubic foot
CIDH	Cast-in-drilled-hole
CJ	Construction joint
CWT	Hundredweight
CY	Cubic yard
DE	Deck edge (square feet)
DF	Douglas fir
EA	Each
F	Cement finisher/or finish
FG	Fine grade
FICA	Federal insurance contributions act (social security)
FIN FLR	Finish floor
F.O.B.	Freight on board
FUI	Federal unemployment insurance
$H5$	Height of form panels—in this case, 5 feet
H_c18	Height to concrete soffit, measured from ground line—in this case, 18 feet
HEx	Hand excavation
HP	Hinge point, or horse power
K	Cost index (see Chapter 2) such as
K_F	Hourly wage rate for 1 cement finisher
K_L	Hourly wage rate for 1 laborer
K_{L+F}	Hourly wage rate for 1 laborer plus 1 cement finisher
K_{6+2}	Total wages for 6 carpenters plus 2 laborers working for 8 hours

K_{Cr}	Hourly rental rate for a 35-ton truck crane
K_{BH}	Hourly rental rate for a small rubber-tired loader-backhoe
L	Length, or laborer, or labor
LS	Lump sum
LCY	Loose cubic yards
LF	Linear feet
M	Materials, or thousand
MB	Machine bolt
MC	Medium curing
MCY	Thousand cubic yards
MFBM	Thousand feet, board measure
M, R	Material or rental costs
MSF	Thousand square feet
MSY	Thousand square yards
NOC	Not otherwise classified
NTS	Not to scale
OD	Outside diameter
OG	Original ground surface
OH	Deck overhang, or overhead
P	Pour or place concrete
PSF	Pounds per square foot
PSI	Pounds per square inch
R	Rentals or rough
RC	Rapid curing
SC	Slow curing
SF	Square feet
SS	Set and strip
SS, P & F	Set, strip, pour, and finish
STA	Stations
SUI	State unemployment insurance
SY	Station yards or square yards
S1E	Surfaced 1 edge
S1S1E	Surfaced one side, one edge
T	Tons (2000 pounds)
t	Thickness
T & G	Tongue and groove
W	Width
WW	Wing walls
Y	Yards
2x, or 4x	Read 2 by, or 4 by. In carpentry, this refers to material which is 2 or 4 inches in thickness

Glossary

Bank yard. A cubic yard of earth, measured in the cut. A cut yard.

Collection hopper. A small hopper used to feed concrete into a wall form or into an elephant trunk.

Cut. The area where excavation is being made, or the volume of excavation, measured in bank or cut yards. The depth of excavation. A decrease in a cost or in a price.

Dead-man. A ground anchorage, usually used in connection with a wire rope bracing system.

Elephant trunk. A tube, 8 or 10 inches in diameter, used to place concrete when the height of fall is greater than about 6 feet. The tube may be made of canvas, rubber, or metal. Elephant trunks are frequently, but incorrectly, called tremie tubes or tremies.

Fill. The area where embankment is being made, or the volume of embankment measured in compacted cubic yards. The height of embankment.

Furring. A system of light wood framing used to enclose ductwork, to lower a portion of a ceiling, or to increase the thickness of a wall, for architectural reasons.

Grizzly. A heavy-duty grid, made of steel shapes or plates, used to screen out oversize material in earthwork or quarrying operations.

Loose yard. A cubic yard of earth, measured after the material has been excavated. Freshly stockpiled earth, and all stockpiled aggregates are normally considered to be in the loose state.

Neat excavation. Excavation made to the neat concrete lines, which are vertical planes at the outermost edges of the concrete.

Sacked finish. A surface finish used to improve the appearance of concrete. The surface is wet thoroughly, and the free moisture on the surface is allowed to dry. Mortar having the consistency of thick cream is applied to the surface with a rubber float. The mortar consists of about ⅔ part portland cement, ⅓ part white cement and 2 parts of well-graded minus 30 mesh sand. The mortar is rubbed over the surface with burlap pads until all air and water pockets are filled. The surface is then dusted with the dry mortar mix, and rubbed again with burlap to achieve a uniform color. On completion, no mortar will appear on the surface, but all voids will be filled. The completed surface is moisture cured for about 3 days.

Scaffold jacks.   Prefabricated metal kneebraces used in connection with temporary staging on wall forms.

Soffit.   The under surface of a slab, a beam, or an arch.

Spandrel beam.   A beam, at the edge of a concrete slab, which extends both above and below the slab.

Spandrel well.   The building wall which extends from the sill of an upper window to the head of the window immediately below.

Starter wall.   A portion of concrete wall, usually 1-½ inches in height, placed as a part of the wall foundation, and subsequently used to spread and align the wall forms.

Tremie tube.   A steel pipe, 10 or 12 inches in diameter with a collection hopper at the upper end, used to place concrete under water.

Wales.   Wooden beams, normally used in pairs to support the wall form studs. Wales are normally placed at right angles to the studs, and are usually made of material which is the same size as the studs. The wales on either side of a formed wall are tied together, through the wall, with form ties.

CONSTRUCTION COST ESTIMATES

1

The Construction Industry

Field experience on construction projects is absolutely essential for those students planning to become general contractors or construction cost estimators. Ideally, this experience should be gained while working in a supervisory capacity for a general contractor. Every construction cost estimator must have the background and ability to function comfortably as superintendent on any job he bids.

This book presents a practical method for preparing detailed cost estimates for construction projects. Since it is written for students who normally will not have had any field experience, this chapter is devoted to the general aspects of the contract procedure and to the elements involved in the bidding and the awarding of construction projects. An understanding of these processes is a necessary prerequisite to the preparation of construction cost estimates, and will also benefit those young engineers and architects engaged in pure design work, who will frequently interface with general contractors during the course of their professional careers.

1.1 Definitions and Terms

An understanding of the following terms is requisite to the discussion of construction cost estimates.

Bare Cost.   The bare cost of a job includes all costs except the contractor's allowance for general overhead and profit. The bare cost of a job item includes the costs for the labor, materials, and equipment required to

1

complete the item, and the labor cost used will include the contractor's allowance for payroll expenses.

Bid Price.   The bid price for a construction project is a firm price for completing the work in accordance with the contract documents. It is submitted to the owner, in writing, by the general contractor. The bid price is determined on the estimate summary sheet and is based on detailed cost estimates. All risks relating to unforeseen job conditions or resulting from estimating errors or omissions are borne by the contractor.

Estimate Sheet.   The sheet on which the estimate of the bare cost of a particular job item is made. Typically, sheets from standard 8½ × 14 columnar pads are used.

Estimate Summary Sheet.   The sheet on which the bare costs for all of the job items are tabulated, together with the costs for direct job overhead, general overhead, and profit. The total of these costs is the bid price for the job. Typically, sheets from standard 8½ × 14 columnar pads having line numbers at the left margin are used.

Item.   The meaning of this term depends on the context in which it is used. A job item is a separate category of work on a particular project. A major item is any category of work which constitutes a large portion of the overall job cost, and may be a job item or a portion of a job item.

Primary Activity.   As used herein, a primary activity is a fundamental operation necessary to complete a portion of a job item. The unit cost for performing a primary activity includes the costs of those secondary activities which must normally be performed to complete the operation. The primary activities are so selected that the relative amounts of these secondary activities will be reasonably constant from one job to another.

Production Rate.   Production rates are expressed in units of work accomplished per hour, for a given crew or piece of equipment. Unit costs may be obtained by dividing the hourly cost of the crew or equipment by the production rate. Production rates determined from past job cost records are said to be "effective rates," since they reflect the inevitable minor delays and interruptions that occur on all jobs.

Rental Rates.   The rental rates for motorized equipment include the costs of operation and maintenance, and the costs of subsistence payments to the operators when required.

Unit Costs.   Unit costs are expressed in dollars per unit of work performed, and are used extensively in the preparation of construction cost estimates. All unit costs must be indexed to the wage or rental rates on which they are based. A production rate may be obtained by dividing the hourly cost of the proposed crew or equipment by the unit cost.

Unit Price.   The unit price is the price bid for completing an item on a unit price contract. The total bid price for a unit price contract is the sum of the products of all the quantities for the various job items and their respective unit prices.

Wage Rates.   The wage rates, to which unit costs for labor are indexed, include the base rate of pay plus all fringe benefits.

1.2   The Estimating Process

Construction cost estimates are made by applying unit costs to the quantities of those primary activities required to complete the various job items. As discussed in Chapter 2, some of these unit costs will be derived from production rates. The soundness of the completed estimate depends on the following three factors:

1. The accuracy of the take-off.
2. The accuracy of the extensions, additions, and transcriptions.
3. The judicious selection of the unit costs and production rates to be used.

1.2.1   Elimination of Errors

Separate take-offs are made for each job item, and the take-offs, in general, follow the same sequence used to perform the item. Each take-off is made to the required degree of precision, and all work is checked. As illustrated in Chapter 4, the primary activities required to complete each item are tabulated and collected in an orderly and systematic manner.

All errors in arithmetic and errors of transcription must be eliminated, since the estimate will be backed by the contractor's money. All extensions are checked and, where the quantities are large, are double checked. All totals are double checked and care is taken to ensure that all figures are clear and legible.

As each step is completed, the work is given a final visual check to eliminate possible errors in magnitude or transcription. The estimate sheets are checked against the take-offs to ensure that no quantities have been overlooked. The estimate summary sheet is checked to ensure that all trades or job items have been listed; and on unit price contracts the quantities listed on the summary sheet for each job item are checked against those listed on the bid proposal.

As the estimate for each job item is completed, the overall unit cost for the item, as estimated, is compared with the overall unit costs realized on past jobs for similar types of work. When significant variation is found, its causes are investigated; this process is simplified when cost records are available to show the overall unit costs for the typical components of similar items of work.

1.2.2   Unfamiliar Minor Items

Estimators are frequently confronted with unfamiliar minor items for which no cost data are available. In such cases, unit costs are reviewed for familiar operations which may or may not be similar to the one in question. A unit cost is selected for a familiar operation that, in the estimator's judgment, will cost more than the one in question, and a second unit cost is selected for an operation that should cost less.

With the probable unit cost for the item in question bracketed, a trial unit cost is selected and converted into a production rate. This rate is considered in view of the anticipated job conditions, is modified if necessary, and the resulting unit cost is entered on the estimate sheet.

On occasion, all estimators face an approaching bid deadline but lack a quoted price for some minor piece of equipment. At any given time, however, the cost per pound is relatively constant for similar types of equipment. A rough estimate can be made in such cases on a weight basis, if the current cost per pound of another piece of somewhat similar equipment is known.

Relative weights are also useful when last-minute quotations for equipment "in lieu of that specified" must be considered. It is always doubtful that a substitution would be permitted where the substitute equipment is appreciably lighter than the equipment called for in the specifications.

1.2.3   The Estimator and the Job Superintendent

Some job superintendents and some crew foremen will do work at lower unit costs than others. The productivity of the proposed job personnel is always considered by the estimator during the unit cost selection process.

Prior to starting a new job, the estimator confers with the superintendent regarding the job procedures and schedules on which the estimate was based. The job superintendent is responsible for completing the work at or below its estimated cost, and in some cases procedural changes may be made. During the course of the work, the estimator monitors the developing unit costs and confers with the superintendent on a regular basis.

1.2.4   Selection of Equipment

All construction cost estimates are based on using particular types of equipment. Normally, equipment selection presents no problems. The contractor will be bidding familiar work in a familiar area and the choice of equipment will be obvious.

On large or unusual jobs, highly specialized and, in some cases, specially built equipment may be required. In such cases the final selection of equipment is made by management, and the production rates for the equipment may be based on the specifications for the equipment together with other information supplied by the manufacturer.

1.3   The Contract Procedure

On every construction project there are but two primary parties or principals involved—the owner, who wants the job done, and the general contractor, who agrees to build it for an agreed price in a specified period of time.

1.3.1   The Job Architect or Engineer

Upon deciding to have a facility built, the owner engages the services of an engineer or architect who prepares the contract documents. Engineers are normally engaged by the owner when the proposed project consists primarily of engineering work such as roads, canals, dams, bridges, and certain industrial facilities. Architects are usually engaged when the proposed project consists primarily of buildings. On most large projects the architect will, in turn, engage the services of a structural engineer, a mechanical engineer, and an electrical engineer, who report to and are paid by the architect.

In all cases the contract documents will consist of a set of plans, the job specifications, the bid proposal, and the contract agreement. In addition, a notice to contractors, contract bonds, and a bidder's bond may be included. While the plans are being prepared, the architect or engineer maintains close contact with the owner and furnishes him with preliminary cost estimates for the developing project. These estimates enable the owner to confirm the economic feasibility of the project, and the plans may be modified to fit the project to the funds available.

Upon his final approval of the plans and specifications, the owner selects a contractor to build the facility. This may be done through negotiations with one or more contractors, or a number of responsible contractors may be invited to bid the project, which will subsequently be awarded to the lowest bidder.

Skepticism is sometimes expressed regarding the concept of dealing with the lowest bidder. Such skepticism may be warranted in those cases where the prices of products or services are compared without regard to the specifications to which the products have been manufactured or to the experience and integrity of the suppliers of the services.

In the construction industry it is the responsibility of the architect or engineer to furnish a complete and detailed set of plans and specifications, a qualified job inspector or resident engineer, and to limit the bidding to competent and responsible contractors. Under such conditions, the awarding of the job to the lowest bidder will enable the owner to obtain the exact job desired at the lowest possible price. From the owner's point of view, his selection of the architect or engineer is of the utmost importance, and he should exercise great care and his best judgment in making this choice.

1.3.2   Notice to Contractors

When the job is to be awarded on the basis of competitive bids, the owner sets a bid date together with a place and time for the bid opening, and the architect or engineer notifies the contractors that the job is out for bids. Public agencies have strict, legislated procedures for notifying contractors, and these procedures must be followed in detail on such work. In practice, all contractors subscribe to daily construction services which publish the essential information concerning jobs out for bid as well as the prices subsequently bid on all projects in any given geographical area.

1.3.3   Prequalification of Bidders

To ensure that all contractors bidding their work are competent and experienced, many public agencies require all prospective bidders to be prequalified. In such cases the contractor files a financial statement and an experience record with the agency on a yearly basis, and is given a rating expressed in terms of the maximum total dollar volume of work he will be allowed to have under construction with the agency during the course of the year. The required bid proposal form will not be issued to any contractor if the estimated cost of the proposed project plus the value of other work he may still have to complete for the same agency exceeds his prequalification limit.

Prequalification of contractors is an excellent requirement from all points of view. It enables the owner to accept the lowest bid with confidence, and it provides an impartial and fair method of screening out prospective bidders who are too small or too inexperienced to bid on a particular project. This screening process thus provides definite benefits to the owner, his engineer or architect, and those contractors who are screened out.

Prequalification should be required more often by architects and engineers in connection with larger private work. The prequalification limit in such cases could be based on a copy of the contractor's current letter of prequalification from the appropriate state agency, and no further analysis by the architect or engineer would be required.

1.3.4   The Cost Estimate

Upon receiving the plans, specifications, and bid proposal, each general contractor prepares a detailed cost estimate for the job. Time is often critical in the preparation of such estimates. The contractor may have only five or six weeks to bid a five story concrete building occupying a city block, for which the concrete take-off alone may require as much as 160 man hours of work.

During the preparation of the cost estimate, the estimator must not spend time analyzing unnecessary details. Each job item must be analyzed only to the degree necessary to determine the lowest probable cost for the job as a whole. Extreme care must be used in estimating the cost of major job items, but time must not be wasted in efforts to be too precise in determining the costs of insignificant items.

1.3.5   Subcontract Costs

In addition to the cost of the general contractor's own work, his cost estimate will include the costs of a number of subcontractors. During the course of construction, the subcontractors report to, are under the general direction of, and are paid by the general contractor. Subcontractors are often referred to by the trade or specialty they perform. Thus the painting subcontractor will frequently be called the painting contractor. The term contractor by itself normally refers to the general contractor. On building projects, about 75 percent of the work is done by subcontractors, while on engineering work the general contractor will do about 65 percent of the work himself.

In all cases the general contractor has the sole responsibility for the successful completion of the project. He furnishes the overall job supervision, prepares all time schedules, and coordinates the work of all of the subcontractors with his own work to ensure that the project is completed in strict accordance with the plans and specifications, in the least amount of time, and at the lowest possible cost to the owner.

1.3.6 The Bid Opening

On the bid day all bids are submitted in sealed envelopes at the time and place designated, and no bids are accepted after the specified time for the bid opening. The bids are normally opened in public at the time specified and are taken "under advisement" for a few days to enable the owner and his architect or engineer to carefully check and consider each proposal. Most specifications allow the owner to reject any or all bids, and all bids are rejected on rare occasions. In the typical case the job is awarded to the lowest bidder.

1.3.7 The Contract Award

Upon being awarded the job, the contractor and his bonding company sign the contract bonds and deliver them to the owner, and the owner and the contractor sign the construction contract. The contractor must then commence work within a specified period of time which frequently will be 5 to 10 working days.

After the contract is signed, the owner is represented by the architect or engineer in all further dealings with the contractor. The contractor does the work in accordance with the plans and specifications and to the satisfaction of the architect or engineer. At the job level, the contractor's superintendent deals on a day-to-day basis with the architect's representative or job inspector, or with the resident engineer in the case of engineering work.

1.3.8 Pay Estimates

During the course of the work the contractor is paid monthly by the owner on the basis of monthly pay estimates which are approved by the architect or engineer. On most jobs the owner retains a specified amount, usually 10 percent of the total amount earned, during the course of the job and for a specified period of time after completion. The retention period after completion is specified to allow mechanics' liens and claims from material or equipment suppliers to surface, if any are forthcoming. In practice, the contractor will receive the final payment about 6 weeks after the owner has accepted the work.

In most cases the contractor receives the monthly payment from the owner 2 to 3 weeks after the monthly pay estimate has been submitted for approval. Because of this time lag and the retention, it is important for the contractor to exercise sound judgment in determining how he will distribute his allowances for direct job overhead, general overhead, and profit among the job items on engineering work, or among the sections of the cost breakdown on building work.

1.3.9   Contract Change Orders

Unforeseen changes in the proposed work, which become necessary or desirable during the course of the job, are specified and paid for on contract change orders. Change orders are negotiated at the job level, approved by the architect or engineer, and paid for at agreed prices when possible. In cases where it is difficult to arrive at an agreed price, the change order work may be done on a time and material, or force account basis. In such cases, the contractor does the required work and is paid for his actual costs plus a percentage for overhead and profit, as set forth in the specifications. If the change order work falls on the critical path for the project, a time extension is negotiated and becomes a part of the contract change order.

1.3.10   Liquidated Damages

Specifications may require the contractor to pay liquidated damages to the owner for each calendar day that the actual time required for completion exceeds the time specified for completion. Such damages may vary from one hundred dollars a day for a simple box culvert job to several thousand dollars a day on large contracts.

Liquidated damages represent an attempt to arrive at a predetermined and agreed price for the actual damages that will be suffered by the owner in the event of delays in completing the work. From the owner's point of view, it is legally important that the specifications do not refer to liquidated damages as a penalty.

When liquidated damages are specified, it is essential that the specified time for completion be reasonable. When, due to unusual circumstances, the time specified for completion is based on using double shifts, this should be set forth in the specifications.

Liquidated damages will not ensure that an impossible time schedule will be met. In such cases the estimated cost of the liquidated damages, based on a realistic time schedule, is added to the estimated bare cost of the job as an added item of expense.

1.4 The Contract Documents

The contract documents must include the plans, specifications, bid proposal, and contract agreement. A notice to contractors, contract bonds, and a bidders bond may be included. All other contract documents are included in the contract agreement by reference, and are an integral part of the construction contract.

1.4.1 Plans and Specifications

On fixed price contracts, the owner must furnish a complete and detailed set of plans and specifications to the contractor. The contractors must be able to determine exactly what is required to be done and to be furnished. Ambiguous or incomplete plans and specifications require the contractors to make assumptions as to what may be required, and often result in higher bid prices from the experienced contractors.

A standard set of specifications is frequently furnished, together with a set of special provisions which relate the general requirements of the "standards" to the particular job by reference. On building work, the specials should include a separate section for each trade involved, and on engineering work the specials should include a separate section for each item of work.

The writer of the special provisions must be thoroughly familiar with the contents and requirements of the standards. Thus it is not sufficient for the specials to specify that the poles for the electroliers be in accordance with the standard specifications when the standards cover the requirements for painted as well as galvanized poles.

When it is desirable to change the requirements of the standards in connection with a given item of work, this is done in the specials by a specific reference. Thus, the specials might state that "in lieu of the method of payment set forth in section 63.07 of the standard specifications, the excavation will be paid for on a lump sum basis." In this case the specials would then say exactly what items were to be included in the lump sum paid.

Care must be exercised not to duplicate any of the requirements of the standards in the special provisions. Such duplication will inevitably result in conflicting requirements which will necessitate the issuance of addenda to the specifications during the time that the job is being bid. Discrepancies between the standards and the specials reflect careless specification writing and should not occur in specifications written by competent engineers or architects.

1.4.2   The Bid Proposal

The bid proposal is addressed to the owner and is signed by the contractor. The contractor inserts his bid prices into blank spaces provided, and no alterations or additions are allowed. A good bid proposal form should not require any price to be entered in more than one place, nor any price to be written in words.

Following the bid opening, all the extensions and additions on unit price contracts are checked by the owner's engineer. Where an extension is incorrect, the unit price is considered to be correct, and the extension, together with the total price bid, is corrected by the engineer to determine the true bid price.

On most building jobs, a list of proposed subcontractors must be submitted with the bid proposal, and the successful bidder can only use the listed subcontractors on the proposed project. During the time that quotations from subcontractors are being received, the list of subcontractors is kept current and in phase with the low quotations being used in the estimate. The requirement for the list of subcontractors results from the efforts of subcontractor associations to combat "bid peddling" by unscrupulous general contractors together with unscrupulous subcontractors. Top quality general contractors maintain strict ethics in this regard, and will never reveal any subcontractor's quoted price.

1.4.2.1   ALTERNATES.   Alternate bids are frequently required on building work, and minor alternates, involving only a few subcontractors or the general contractor's own work, can be processed accurately by the contractor on bid day. Such an alternate might be one requiring the additional cost for using a different type of lighting fixture in certain rooms in lieu of the ones specified in the base bid, or the reduction in cost if certain planters specified in the base bid are deleted from the contract.

There is insufficient time available on bid day to accurately process a major alternate involving many or all of the subcontractors. The reasons for this shortage of time are inherent in the bidding process, and are completely beyond the control of the general contractor. The problems involved are discussed in Section 1.9, which covers the procedure on bid day.

Bid proposals requiring major alternates will be submitted by the contractors, since their livelihoods depend on their obtaining work, but as a rule the cost to the owner for such work will be greater than if the alternate were included as a part of the base bid. On building work, the spread between the base bids submitted by the first and second bidders may be 1 percent or less, while the spread between their major alternates frequently amounts to 20 percent or more.

Major alternates often cause the listed subcontractors to change, depending on which combination of base bid plus alternates is selected by the owner. For the same reason, even the low general contractor may be indeterminate and the owner will, in effect, have two low bidders. In such cases, the owner and the architect will have at least one unhappy contractor regardless of what decision they make.

The failure of such a proposal to present the owner with a single undisputed low bidder may result in considerable embarrassment and disadvantage to the owner in cases where other business or political considerations are involved. A competent architect performs a very real service for the owner when the need for major alternates is eliminated through proper use of preliminary cost estimates and discussions during the planning stage of the project.

1.4.3   The Bidder's Bond

The bidder's bond is a guarantee that the contractor will enter into the contract with the owner at his bid price if he is awarded the job. Such bonds are addressed to the owner, signed by the contractor and by his bonding company, and are normally in the amount of 10 percent of the amount bid. Should the contractor refuse to enter into the contract he could be required to forefeit the amount specified on the bond to the owner.

The cost of the bidder's bond is nominal, and is a general overhead item. Many bonding companies make no charge for supplying bidder's bonds to those contractors for whom they regularly furnish contract bonds.

1.4.4   The Contract Bonds

Two contract bonds are normally required. The labor and materials bond is a guarantee that the contractor will pay his workmen and his suppliers. The performance bond is a guarantee that the contractor will complete the work in accordance with the requirements of the plans, specifications, and contract. These bonds are addressed to the owner and are signed by the general contractor and by his bonding company.

In the event of default by the contractor, the bonding company will have the work completed, usually by another contractor. The bonding company will pay all outstanding bills, which are subject to lien laws, with the exception of insurance premiums and fees for accountants and attorneys. The bonding company will recover its costs by liquidating the contractor's business and personal assets, if necessary.

From the owner's point of view, it is essential that the specifications clearly define what conditions will constitute default by the contractor, and what will be grounds for terminating the contract. In addition, the procedures and time periods to be followed and allowed in the event of default should be set forth clearly.

Some architects and engineers have experienced difficulty in getting a bonding company to take over a job on which the contractor was not performing. In most cases, such difficulties result from specifications which fail to precisely define the grounds and the procedures for terminating the contract.

The cost for the two contract bonds is obtained from the bonding company, and the cost will vary with the size and classification of the job. Current bond costs are:

Job price	Cost per thousand dollars of job price	
	Class A	Class B
First $ 500,000	$9.00	$12.00
Next $2,000,000	5.60	7.25
Next $2,500,000	4.40	5.75
Next $2,500,000	4.10	5.25
Over $7,500,000	3.70	4.80

The preceding rates apply to jobs requiring up to 24 months to complete. A surcharge of 1 percent of the regular premium is made for each month of job duration in excess of 24 months. The minimum bond premium is currently $25.00.

Class A work includes airport runways, bridges, culverts, curbs, gutters, sidewalks, grading, highways, paving, and underpasses. Class B work includes buildings, canals, dams, embankments, piers, piling, retaining walls, sewage plants, subways, and tunnels.

1.5 Types of Estimates

The term "estimate" is loosely used in the construction industry, and its precise meaning must often be determined from the context in which it is used. It may refer to detailed cost estimates, preliminary estimates, estimates of the quantities involved in the job items, or estimates of moneys due to the contractor in payment for work done.

1.5.1   Detailed Cost Estimates

These are the estimates prepared by contractors in bidding jobs, and such estimates are covered in depth in the following chapters.

All detailed cost estimates are based on a take-off of the materials and of the primary activities required for each job item, and the quality of the completed estimate depends on the accuracy of these take-offs.

The total cost of any project consists of four categories of expenses. These are the costs of those materials to be furnished by the general contractor, the costs of subcontractors, the general job costs, and the costs of the labor and equipment required to do the general contractor's own work. The costs of materials and the subcontract costs are based on fixed quoted prices, and can be determined accurately. Although sound judgment is required, the general job costs can be estimated with a reasonable degree of accuracy. The area of greatest uncertainty in the preparation of detailed cost estimates lies in the selection of the proper unit costs or production rates for labor and for equipment in connection with the general contractor's own work.

The total labor and equipment costs for the general contractor's own job items consist of the sum of the products of many quantities and their respective unit costs. Although the majority of the unit costs will be about right, some will be too high and others will be too low. In practice the sum of these products, however, will be very close to the costs actually realized in those cases where experienced estimators use sound judgment in selecting unit costs based on past cost records. Extreme care must always be used in selecting unit costs where large quantities are involved. The two primary pitfalls to be avoided are the failure of an inexperienced estimator to recognize some essential activity and to overlook it entirely, or where, in the case of unit costs based directly on production rates, some circumstance on the particular job that will greatly affect the rate of production goes unnoticed.

1.5.2   Preliminary Cost Estimates

Preliminary estimates are frequently referred to as architect's estimates or as engineer's estimates. Architect's estimates are often based on the total square feet of floor area or on the total cubic feet of building volume for a given type of building and a given class of building use. There are publications available that present such data together with the realized unit costs for buildings recently bid in various sections of the country. Such data are often used to supplement the architect's own cost records, based on his own work in his own area.

Engineering projects are normally bid on a unit price basis. Daily construction services publish the results of all public bid openings in their areas and a file of average unit prices for the various items of work can readily be built up. Contractors often make estimates based on such average unit prices and a published engineer's estimate of quantities, prior to having any plans or specifications, as an aid in determining whether or not to bid a particular project.

Preliminary estimates can be made quickly, but are not sufficiently accurate to use as the basis for a bid. The architect or engineer may, to some degree, risk his reputation with a preliminary estimate, but the contractor is risking money.

1.5.3   Quantities Estimates

The list of job items together with their estimated quantities is called "the engineer's estimate," and constitutes the major portion of the bid proposal form on engineering projects. In addition, the engineer's estimate should be the first page of the special provisions for engineering work.

The quantities shown on the engineer's estimate are for estimating purposes only, and the final pay quantities are based on calculations using plan dimensions supplemented by field measurements when necessary. The pay limits used in preparing the engineer's estimate are set forth in the specifications, and these are the limits used in determining pay quantities. The specified pay limits may be different from those required to do the work. The horizontal pay limits for structure excavation and structure backfill are frequently specified to be vertical planes, and the contractor may actually move and replace more than twice the number of cubic yards shown on the engineer's estimate.

The number of items on the engineer's estimate may vary from 4 or 5 in the case of a simple box culvert job, to as many as 150 or 200 on a 15 mile stretch of freeway.

1.5.4   Pay Estimates

Pay estimates are estimates of the money due to, and earned by, the contractor. Monthly pay estimates are normally made during the course of the work and upon completion of the job. After completion and upon the expiration of the specified waiting period, the final payment is made, based on the final pay estimate. The final payment consists primarily of the money retained during the course of the work, and in the case of unit price bids is based on a final and precise determination of the total quantities of all of the job items.

On building work, which is bid on a lump sum basis, the contractor furnishes the architect a cost breakdown by job sections at the start of the work. Subsequent pay estimates are based on an estimated percentage of completion for the items on the cost breakdown.

1.6   Types of Construction Contracts

Negotiated contracts are frequently awarded on a cost-plus basis, while competitively bid contracts are always awarded on the basis of a fixed price for doing a specified job.

1.6.1   Cost-Plus Contracts

Cost-plus contracts are typically used in those cases where complete plans and specifications are not available at the time the owner must start the project. Under the cost-plus system, the contractor is paid for his actual job costs plus a percentage or a fixed fee for general overhead and profit. In such cases the contract must specify exactly what labor and equipment is to be paid for, the rental rates for the equipment, and what labor and equipment are to be included in the contractor's general overhead. Extensive records must be kept at the job and at the home office covering the cost of materials used, the job payrolls, and the times of use for equipment. These records are kept on a daily basis, certified by the contractor, and checked and approved by the job inspector or by the resident engineer.

On cost-plus work, the owner saves some of the engineering costs otherwise required to prepare a complete set of contract documents and, more importantly in many cases, saves the time required to prepare the complete set of documents and to hold a formal bid opening. In addition, on negotiated contracts, the owner is able to select a particular contractor to do the work, and in some cases business considerations may cause this to be an important feature from the owner's point of view.

On the other hand, cost-plus work normally costs more than competitively bid work. There is less incentive for the contractor to keep costs down, more record keeping is involved, the owner pays for all unforeseen problems, and, in many cases, the contractor charges the owner more for his own equipment than he would charge himself if he were bidding the work on a competitive basis.

An additional and serious disadvantage of the cost-plus system is that it does not enable the owner to determine the final cost of the proposed project in advance.

1.6.2   Fixed Price Contracts

Fixed price contracts may be bid as a lump sum for the entire job, or on the basis of unit prices applied to the engineer's estimate of the quantities involved for each of the various job items.

1.6.2.1   LUMP SUM CONTRACTS.   Buildings are bid on a lump sum basis. In nearly all cases the total quantity of materials and work to be done can be precisely determined from the plans and specifications.

On lump sum bids, the estimator lists the job items on the estimate summary sheet. The listed items correspond to, and are in the same sequence as, the sections covered by the specifications.

There may be sixty or more sections involved, most of which cover specific subcontract work. The lump sum price bid is the sum of the bare cost of the contractor's own work including his direct job overhead, the subcontractors' prices for their respective sections, and the contractor's allowance for general overhead and profit.

Lump sum contracts require the contractor to construct a complete facility in accordance with the plans and specifications. All required work and materials must be included in the estimate, even though some may not have been covered in the sections of the specifications.

1.6.2.2   UNIT PRICE CONTRACTS.   Most engineering work is bid on a unit price basis. On such work it is usually impossible to determine the final quantities for some items from even the best of plans, and the unit price method enables the contractor to receive payment for the actual quantities of work done based upon field measurements made during the course of construction. An important secondary benefit derived from unit price bidding is the relative ease of making preliminary estimates for future work, based upon a file of published unit prices bid on similar past jobs.

On unit price bids, the estimator lists the job items and their quantities on the estimate summary sheet. The list is identical to the bid proposal and to the engineer's estimate. The unit prices bid will include the contractor's costs for direct job overhead, general overhead, and profit. Only that work covered by the job items, or specified as included in the cost of one or more of the job items, need be done.

The unit *prices* bid for items on the engineer's estimate should not be confused with the unit *costs* applied to the various activities required to complete the items and which are the basis of the cost estimate. In the broadest sense, the unit price bid for any item is determined by dividing its total estimated cost by the number of units on the engineer's estimate for the item.

On most unit price contracts, some of the items themselves are bid on a lump sum basis. Lump sum items save engineering time during the preparation of pay estimates, and the contractor saves time during the bidding process, since it is not necessary to check the engineer's estimated quantities as must be done when the unit price method is used and the pay limits differ from those required to do the work.

Overhead and profit could theoretically be prorated to the bare costs of the various items of work. This is not practical because of lack of time on bid day and uncertainties regarding the final pay quantities of some of the items on the engineer's estimate. The contractor cannot afford to assign appreciable amounts of overhead and profit, which are fixed costs, to an item that may under-run by a substantial amount on completion.

In practice, therefore, overhead and profit are arbitrarily distributed to the major items of the contractor's own work, and in particular to items bid as a lump sum, to items whose final pay quantities, as shown on the engineer's estimate, are known to be reasonably correct, and to items which may be expected to over-run the engineer's estimate. Early pay items are favored over late pay items, since this enables the contractor to start generating working capital at an earlier date.

1.7   Subcontractors

The subcontractor is a specialist in a particular field. He exists because he can do his special work for less money than the general contractor would spend, overall, if he were to do the work himself. The subcontractor furnishes all labor, materials, and equipment necessary to complete a portion of the project. His completed work must conform to the plans and specifications in all respects, but his method of doing his work must be entirely under his own control. He works for and is paid by the general contractor.

Subcontractors make their own take-offs, quote their work at a fixed price, and furnish their own direct job supervision. As a result, the time required by the general contractor to bid work, his general overhead and job supervision costs, and the amount of uncertainty at bid time are all reduced. The subcontractors, as a group, furnish the general contractor with most of the advantages of a greatly increased organization at no additional cost to the contractor or the owner.

Essentially, all subcontract quotations are made by telephone on the bid day and are subsequently confirmed in writing to the low bidder. On building work, subcontractors normally quote a complete section of the special provisions, and in such cases their quotations are noted to be "plans and specs." Special provisions for engineering work cover the work by job items and not by sections relating to subcontract trades. On such work, subcontractors frequently quote only a portion of a particular job item.

1.7.1 The Subcontract Agreement

On being awarded the job, the general contractor enters into subcontract agreements with those subcontractors whose quotations were the lowest at bid time. At the very minimum, the subcontract agreement should clearly specify the work to be done and the method of payment, should bind the subcontractor to the general contractor in the same manner that the general contractor is bound to the owner by the contract documents, and should specify the procedures and time periods to be followed and allowed in case default by the subcontractor requires his removal from the job. On building work, in particular, subcontractors should be required to clean up their own work during the course of the job and upon completion of their work.

1.7.2 Subcontractor's Bonds

The subcontractor's bond is a guarantee that the subcontractor will do the work and pay his labor costs and his suppliers. It is addressed to the general contractor and is signed by the subcontractor and by his bonding company.

The cost of a subcontractor's bond is essentially the same as the cost of the usual contract bonds but is based on the amount of the subcontract. When subcontract bonds are required their cost is added to the subcontractor's quotation.

1.7.3 Material Suppliers

On building work, a number of major material suppliers do their own take-offs and quote a fixed price by telephone on bid day. They are similar to subcontractors, but they do not include the installation of their materials. Among such suppliers are those who furnish hollow metal doors and frames, millwork, wooden doors, and finish hardware. In all cases their quotations should be "plans and specs" for the material furnished.

1.8 Joint Ventures

Two or more contractors may join forces and form a partnership or joint venture for purposes of bidding and constructing a particular project.

Such a joint venture is a separate legal entity, has its own name, address, and required licenses, and the contract bonds are made in its name.

Joint ventures are sometimes formed when one or more of the partners is unable to bid a proposed project alone because of prequalification or bonding company limitations.

Specifications for engineering work frequently require the general contractor to do 50 percent or more of the work with his own forces. When a proposed project is fairly evenly divided between two primary categories of work, it may be impossible for a contractor specializing in one of them to do 50 percent of the work if he subcontracts the other. In such cases, a joint venture enables two contractors, each an expert in one of the major work categories, to submit a joint venture bid that is lower than would be probable if each were to bid the job alone.

Since a joint venture is a partnership, each partner is responsible to the owner for the successful completion of the project, and it is therefore imperative that each of the joint venturers have complete confidence in the integrity, ability, and financial strength of the others. Should one or more of the joint venturers go bankrupt or fail to perform, the remaining venturers must complete the project regardless of their particular areas of expertise.

Although all joint ventures are treated as separate legal entities by the owner, they may be structured internally either as true joint ventures or as item joint ventures.

1.8.1   True Joint Ventures

In a true joint venture the partners contribute working capital to the joint venture bank account, hire all office and job personnel on the joint venture payroll, buy all supplies, and rent or purchase all equipment in the name of the joint venture. In addition, all invoices and payrolls are paid with checks drawn on the joint venture bank account. On completion of the work the profit made by the joint venture is distributed to the partners as set forth in their joint venture agreement.

1.8.2   Item Joint Ventures

On an item joint venture, each of the venturing partners is responsible to the joint venture for a specific number of job items set forth in the joint venture agreement. The prices each partner quotes to the joint venture for his items include his overhead and profit, and his items normally include those subcontract items relating to his area of work.

One of the venturers acts as the managing partner for the job, and his address is used as the joint venture address. The managing partner combines the prices quoted by the other partners with his prices for his own items and submits the bid.

If the bid is successful, a bank account is set up in the joint venture name for the sole purpose of processing payments from the owner. Such payments are deposited therein and are immediately disbursed to the partners, in accordance with their earnings on their items of work.

On item joint ventures, each partner hires all personnel for his items of work on his payroll and pays all of his suppliers with checks drawn on his bank account. In the broadest sense, the managing partner performs as the general contractor for the job, and the other partners perform as subcontractors.

Upon completion of the work, the amount of profit realized by each venturer will depend on his success in bidding and completing his items of work. On an item joint venture, it is thus possible for one or more of the partners to make more than the estimated amount of profit, while the others may make less.

1.9 Procedure on Bid Day

The actual procedures followed by different contractors on bid day will vary, but the basic operations required and the conditions that exist are the same for all of them. An appreciation of the conditions confronting the contractor on bid day is essential to the estimator and to those engineers and architects who prepare contract documents and hold bid openings. The hectic situation that exists when large building jobs are bid results from conditions that are beyond the control of the contractor. Bid proposals requiring unessential information or many addenda add to the confusion and result in higher bid prices.

The contractor should have all his own work completed and entered on the summary sheet prior to bid day. The sheets used are of legal size and are taken from standard columnar ruled pads.

1.9.1 Bidding Building Work

The majority of the subcontractors submit their quotations by telephone during the last two hours prior to the bid opening. It is not unusual to receive as many as twenty-five quotations or cuts in prices during the last half hour. The contractor must handle this telephone traffic, enter and revise quotations on the summary sheet, revise the list of subcontractors and keep it in phase with the changing quotations, make a final determination of his allowance for general overhead and profit, determine the total job cost, fill out the bid proposal, and deliver the sealed bid to the designated location prior to the bid deadline. In many instances, the place designated for the bid opening will be several hours away from the contractor's office where the bid is being prepared.

A supply of printed forms on which subcontractors' bids are entered should be available together with a suitable file box for them. In addition to having spaces for the quotation itself, these forms should have a block at the top where the line number of the section quoted, as set up on the estimate summary sheet, may be entered. The quotations are filed numerically by line numbers. Since the contractor must frequently refer to previous quotations, it is imperative that he be able to retrieve them with no unnecessary loss of time.

A typical estimate summary sheet for a reinforced concrete building project is shown in Figures 1.1 and 1.2. The lines are numbered consecutively at the lefthand margin. All costs are entered in column 1, and columns 2 and 3 are used for processing last minute cuts in prices.

The contractor enters his own name on line 1, together with the total cost for all of his own work; this total includes his cost for direct job overhead.

The estimated cost of the contract bonds is entered on line 2, together with a notation of the total estimated job cost on which the bond cost is based and the rate of change in the bond cost per thousand dollars of bid price.

On the project illustrated, the cost for fire insurance was paid by the owner, and the liability insurances were based on a composite rate and assessed as a part of the payroll expenses, which are discussed in Chapter 3. When applicable, insurance costs normally are entered immediately following the cost of the bonds, although some contractors carry insurance costs on their direct job overhead sheet.

In our example, starting with line 3, the sections from the specifications that are to be subcontracted are listed in the same sequence in which they are covered in the specifications. Those sections relating to the general contractor's own work are omitted, since all of his costs were combined and entered on line 1.

A separate sheet on which the names and addresses of the successful (low) subcontractors will be listed, is set up in the same order as the estimate summary sheet. This sheet is called "the subcontractors list" and the line numbers for each trade listed are identical to those on the summary sheet. The subcontractors list will not show any of the prices quoted.

By keying the quotation forms and the subcontractors list to the line

SUMMARY - CITY HOSPITAL

LINE NO.	SEC.		COST	REVISED COST	ADD CUT
1		ABC CO.	1,371,000		
2		BONDS (7500M, +4.80, -5.25)	48,000	50,000	+2,000 ✓
3	2A	EARTHWORK	88,500		
4	2B	A.C. PAVING	3,710		
5	3A	REINF. STEEL	345,000		
6	3B	PRECAST CONC.	69,800		
7	4A	MASONRY	5,910		
8	5A	STRUCTURAL STEEL	117,900		
9	5B	ARCH'L. METAL	44,000		
10	6B	MILLWORK, WOOD DOORS	74,500		
11	7A	MEMBRANE ROOFING	47,200		
12	7B	B'LD'G. INSULATION	16,900		
13	7C	SPRAYED INSULATION	13,500		
14	7D	SEALANTS	3,750		
15	7E	DAMPPROOFING	22,400		
16	7F	AGGRE-DECK	14,500		
17	7G	FLEX-DECK	17,200		
18	8A	STORE FRONT			
19	8B	ALUM. WINDOWS			
20	8C	PORCELAIN PANELS	149,500		
21	8D	ALUM. DOORS			
22	8E	GLASS, GLAZING			
23	8F	AUTOMATIC DOORS	21,000		
24	8G	H. M. DOORS, FRAMES	49,100	44,000	- 5,100 ✓
25	8H	FOLDING DOORS	10,500		
26	8J	FINISH HARDWARE	58,500	57,500	- 1,000 ✓
27	8K-2	VAULT DOOR	1,250		
28	8K-3	ROLLUP DOORS	1,650		
29	8K-4	FIRE DOORS	5,200		
30	9A	LATH & PLASTER	532,500	491,000	- 41,500 ✓
31	9E	ACCOUSTICAL	28,500		
32	9F	PAINTING	70,000		
33	9G	CERAMIC TILE	59,500		
34	9H	TERRAZZO	49,400	52,000	+ 2,600 ✓
35	10A	TACK, CHALK BOARD	430		
36	10B	TOILET COMPARTM'TS.	2,300		
37	10C	LINEN CHUTES	6,100		
38	10D	METAL LOCKERS	6,950		
39	10E	TOILET ACCESSORIES	26,700		
40	10F	CUBICLE TRACKS	12,600		
41	10G	NAME PLATES	2,600		

Figure 1.1 Estimate summary sheet.

23

LINE NO.	SEC.		COST	REVISED COST	ADD CUT
		SUMMARY – CITY HOSPITAL (CONT'D.)			
42	10H	INCINERATOR	3,920		
43	11A	KITCHEN EQUIPM'T.	69,900		
44	11B	MODULAR MET. CASEWORK	253,000		
45	11C	AUTOPSY EQUIPM'T.	10,500		
46	11D	JANITOR STATION	7,350		
47	11E	CHART·HOLDER	2,890		
48	11F	FILM ILLUMINAT'N.	4,560		
49	11G	KITCHEN UNIT	700		
50	12A	DRAPERY TRACK	3,450		
51	12B	CARPET, RESILIENT FLOOR	103,900	101,000	−2,900
52	12C	VERTICAL BLINDS	8,400		
53	13A	RADIATION PROTECT'N.	11,800		
54	13B	MOVABLE WALL SYST.	112,500		
55	14A	PASS. ELEVATORS	211,200		
56	14B	FREIGHT ELEV. (EST)	30,000	27,000	−3,000
57	15A	SHEET METAL			
58	15B	HEAT'G. & VENTILAT'G.			
59	15C	PLUMBING	2,319,600		
60	15D	MEDICAL GASES			
61	15E	FIRE SPRINKLERS	28,500		
62	16A	ELECTRICAL	1,061,700	1,020,000	−41,700
63			—		
64				CUT	−90,600
65		TOTAL LESS [(3)]	5,142,820		
66		TOTAL [(3)]	2,499,100		
67		TOTAL INCL. [(3)]	7,641,920		
68		G.O. + PROFIT	350,000		
69		TOTAL	7,991,920		
70		CUT	90,920		
71		BID PRICE	7,901,000		

Figure 1.2 Estimate summary sheet (continued).

numbers from the estimate summary sheet, the required entries and their revisions can be made in a minimum amount of time.

From three to six people will be required to take quotations over the telephones. The estimator has the original copy of the summary sheet and the others have duplicate copies. These duplicates are used to key quotations to line numbers, and no prices are entered on them.

Most bid openings are held in the afternoon. At the start of the day, the quotations from subcontractors come in slowly. The quotations are entered on the forms, and the line number for the section quoted is entered in the block provided. Each quotation is given to the estimator, who enters it in pencil on the summary sheet, lists the subcontractor, and files the quotation form numerically by the line number. During the final two hours prior to the bid opening, another person will be required to process the subcontractors list, and this person should be within arm's reach of the estimator.

As subsequent quotations are received, the estimator compares them with those previously entered on the summary sheet, erasing the previous entries and inserting the new ones if they are lower. The subcontractors list is revised to correspond, and the quotation forms are filed by line number, with the low quotation in front.

Forty-five minutes prior to bid opening, all telephones will be ringing constantly, and there will normally be 3 or 4 sections for which no quotations have been received. The estimator encircles these open spaces in the cost column to make them easily identifiable, and makes a duplicate copy of his summary sheet. In our example these openings are centered on lines 20, 56, and 59. At the time the duplicate copy is made, most of the quotations listed in column 1 will have been revised from 1 to 4 or more times. The total for column 1, less the encircled openings, is entered on the estimator's duplicate summary sheet; in our example this total is shown on line 65. This procedure enables most of the addition to be made and checked prior to the last minute rush.

From this point on, the estimator works with the duplicate summary sheet, and there is literally no time to spare. No further erasures are made in column 1, and subsequent changes in prices are processed in columns 2 and 3. In our example, last minute revisions are shown on lines 2, 24, 26, 30, 34, 51, 56, and 62.

During this time, the estimator must normally contact one or more subcontractors for clarification of their quotations. In addition, combined quotations for several sections may be received, and must be compared with the separate quotations for the same sections. This comparison is best made on a separate worksheet, as discussed below.

When the designated location for the bid opening requires more than 10 minutes of travel from the contractor's office, a satellite employee, with the signed bid proposal and subcontractors list, is positioned at a telephone near the designated location. In those cases where the satellite employee is located several hours away from the home office, he will be in constant communication with the office during the last half hour. He is given the list of subcontractors as it then stands, makes the listing in pencil, and revises it as the subcontractors change. Since the list is keyed to the line numbers on the estimate summary sheet, this procedure is fast and there is little chance for error.

Twenty minutes prior to bid time, all openings in column 1 of the summary sheet will normally be filled, but all telephones will be ringing, as the subcontractors cut their earlier quotations. In those cases where an encircled opening has not yet been quoted, the estimator inserts an estimated value for the item, and determines the sum of all of the encircled quantities. Thus, in our example, an estimated price of $30,000 is entered in the opening at line 56, and this figure is corrected, when the quote is received, in columns 2 and 3.

The total for the encircled quotations is entered (line 66 in our example), and added to the previous total. The contractor's final allowance for general overhead and profit (discussed in Chapter 3) is finalized and added to the previous total to arrive at a bid price which is approximate, since the cuts in column 3 of the summary sheet have not been made.

The cost for the contract bonds is revised to correspond to the approximate bid price, and the change in cost is entered in column 3 of the summary sheet. In the example this operation is shown on line 2.

The additions and cuts shown in column 3 are added algebraically, and the resulting cut is deducted from the approximate bid price to obtain the final bid price for the job. In the example this process appears on lines 70 and 71. In a typical case, the cut will be made about 10 minutes prior to bid time, and the telephone traffic will have decreased. When additional last-minute cuts are received, they are processed in columns 2 and 3 of the summary sheet, using a colored pencil for easy identification. The bid price is revised to correspond, and is given to the satellite employee, who enters the bid price, in ink, on the proposal, makes the final revision to his subcontractors list, seals the proposal in a pre-addressed envelope, and delivers it to the designated loation prior to the designated time for the bid opening.

Alternates, when required, are usually processed on separate summary sheets for each alternate. On these sheets, only those sections relating to the alternates are listed. Simple alternates, involving only 2 or 3 trades, are sometimes processed on the estimate summary sheet, which will require three additional columns for each alternate.

1.9.1.1 PROCESSING COMBINED QUOTATIONS. During the last hectic hour, the contractor is frequently confronted with combined quotations for two or more sections which are also being quoted separately by other subcontractors. To avoid confusion and error, he must be prepared to evaluate such combinations in a systematic and orderly manner, and the procedure should be such that the comparison can be made by a qualified assistant.

The comparisons are made on a separate sheet, such as the one shown in Figure 1.3, and the format for this comparison sheet should be set up prior to the day of the bid opening. Usually 3 job sections, or trades, will be involved.

On the comparison sheet shown in Figure 1.3, the trades quoted separately and in combinations are listed on line 1, together with their respective line numbers from the estimate summary sheet. Line 2 shows the separate quotations for the three sections as entered on the estimate summary sheet. The total for the three sections is listed in column E. The names of the subcontractors submitting these quotations are entered on the subcontractors list at the time the quotations are entered on the summary sheet. After the quotations have been entered on line 2 of the comparison sheet, no changes, erasures, or comparisons of these sections are made on the summary sheet, and no changes are made in line 2 of the comparison sheet.

Comparison sheet lines 3, 4, and 5 show the low quotations and the

	SECTION (LINE NO.)	COMBINED QUOTATIONS				
		A	B	C	D	E
1	SECTION (LINE NO.)	DRYWALL (32)	ACCOUSTIC. (34)	LATH & PL. (39)	TOTAL	TOTAL
		ENTERED IN COST COLUMN				
2		53,000	31,000	310,000		394,000
	SUBCONT'R.	LOW QUOTATIONS				
3	JONES	52,000				
4	BROWN		30,000			
5	SMITH			300,000	382,000	
6	AJAX	80,000	INCL.	—	380,000	
7	KELLEY	—	INCL.	327,000	379,000	
8	ACME	INCL.	INCL.	INCL.	375,000	375,000

CUT 19,000

Figure 1.3 Comparison sheet for combined quotations.

subcontractors names for the separate quotations subsequently received. These data may be revised as necessary. The total for the three separate low quotations is entered on line 5, column D.

Lines 6, 7, and 8 show the names and the quotations for the low subcontractors submitting combined quotations. Thus, Ajax quotes $80,000 for drywall plus accoustical, which is added to the $300,000 shown on line 5 to arrive at a total for the three sections, which is entered on line 6, column D.

Similarly, Kelley, line 7, quotes $327,000 for lath and plaster plus acoustical, which is added to the $52,000 shown on line 3 to arrive at a total for the three sections, which is entered on line 7, column D.

Acme quotes all three sections for a total of $375,000, which is entered on line 8, column D. The data entered on lines 6, 7, and 8 may be revised as necessary when lower quotations are received.

The totals shown in Column D, lines 5 through 8, represent the low combinations for the three sections, and the lowest one, on line 8, is entered in Column E. The $19,000 cut is the difference between this figure and the $394,000 total for the three sections as entered on the estimate summary sheet. At the time this cut is transferred to the estimate summary sheet, the subcontractors list is revised to show Acme as the subcontractor for lines 32, 34, and 39.

1.9.2   Bidding Engineering Work

On engineering work, fewer subcontractors are involved, the telephone traffic is relatively light, no subcontractors list is required, and there are normally no alternates. On the other hand, the estimate summary sheet and the bid proposal may contain a hundred or more job items, and consequently require more time to complete.

The bid proposal form contains a consecutively numbered list of the job items and their estimated quantities. Spaces are provided for listing the unit price and the total price bid for each job item. The job price is the sum of the prices bid for the listed job items.

The estimate summary sheet is similar to the bid proposal, except that it has an additional column for listing the bare cost of each item. The contractor's direct job overhead cost and his allowance for general overhead and profit are listed in this cost column below the last job item. Since the proposal does not include items for overhead or profit, these costs must be distributed among, and included in, the prices bid for the various job items. On completion, the total for the cost column will equal the total for the total price bid column.

Typical estimate summary sheets for small projects are shown in Figures 4.8 and 4.15. Sheets from standard 8½ × 14, 4-column pads are normally used. The list of job items is double spaced, and is identical to the one on the bid proposal. It is imperative that the quantities listed on the summary be checked against those shown on the proposal, since all extensions are made and checked on the summary prior to being entered on the proposal itself.

On the day before the bid opening, the contractor lists the bare cost for each of his items and his direct job overhead cost on the summary sheet. Some items will be partially subcontracted, and the subcontract costs, when received, will be entered in the double space below the item involved. Horizontal lines are drawn through all unneeded double spaces in the cost column and through all double spaces in the total price bid column, to ensure that all required spaces in both columns will be visible and ultimately filled. On completion, these two columns will have a number or a line in every space.

The unit prices for all but the major items of the contractor's own work are determined by dividing their bare costs by the number of units involved. The results are rounded off, entered, and extended on the summary sheet.

Estimated unit prices for minor subcontract items, together with their extensions, are entered on the summary sheet, even though the actual quotations will not be received or entered in the cost column until the following day. With the summary sheet thus filled out as far as possible, the available prices are entered on the bid proposal in ink. As each item is entered on the proposal, a hashmark is drawn through the checkmark on the corresponding total price on the summary sheet. This enables the contractor to determine which items have been entered on the proposal by scanning the summary sheet, and is important in those cases where the signed proposal will be in the hands of a satellite employee on bid day.

On bid day, the subcontractors will quote on a unit price basis. These unit prices are multiplied by the quantities involved, and the resulting costs are entered in the cost column of the summary sheet. As a general rule, the unit prices quoted are rounded off and entered, together with their extensions, in the price columns of the summary sheet. Where large subcontracts are involved, the contractor may assign a portion of his costs for overhead and profit to the item.

The costs for the contract bonds and special insurance are normally carried on the direct job overhead sheet, on unit price bids. Revisions to these costs and last minute changes in subcontractors' quotations are processed on a separate cut sheet, as discussed below.

Ninety minutes before the bid opening on larger jobs, there may be four or five subcontract items that have not been quoted, and will be open in the cost column of the summary sheet. These same items together with two or three of the contractor's own major items will also be open in the total price column. The estimator encircles the open items in each column to ensure their easy identification and enters the totals for each column, less its encircled openings, on the summary sheet. The total for the cost column includes the cost of the contractor's direct job overhead. The above procedure enables the contractor to complete and check the bulk of the required addition prior to the last minute rush.

In cases where a satellite employee has the bid proposal at a distant location, he will, by this time, have the proposal filled in except for the open items, and he will keep the bid proposal current from this time on.

Forty minutes prior to bid time, all open spaces in the cost column will be filled. Their sum, plus the cut from the cut sheet, are added to the preceding total to obtain the bare cost of the job at that time. The allowance for general overhead and profit is finalized and added to the bare cost to obtain a preliminary total job cost, which will be revised as later cuts are received.

The remaining unit prices are established as discussed previously in Section 1.6.2.2, but with the added criterion that the total price bid must equal the total job cost. The unit prices for all but one of the open items are entered on the summary sheet and the proposal, together with their extensions. The sum of these prices is added to the previous total for the total price column to obtain the total bid price for the job, less the bid price for the one open item. During this period, additional cuts are being processed on the cut sheet, using a colored pencil, and the cuts are listed in a second column designated "cut number 2."

Fifteen minutes before bid time, cut number 2 is totaled, and the final job cost is determined. The difference between this cost and the last total for the price column represents the total price for the open item, which will exactly balance the bid price with the total job cost. In practice, the resulting unit price for the open item is rounded off, and the total price for the item, based on the rounded off unit price, is added to the previous total to obtain the total bid price for the job. The prices for the open item and the total job price are entered on the bid proposal, which is sealed and delivered to the designated location prior to the designated time for the bid opening.

1.9.2.1 THE CUT SHEET.   Unit price jobs require an estimate summary sheet which is considerably more complicated than the ones used for lump sum bids. To avoid confusion, late revisions to subcontractors' quotations, material prices, and certain direct job overhead items are processed on a separate cut sheet.

CUT SHEET					
1	2	3	4	5	6
NO.	ITEM AND UNIT COST	ENTERED	REVISED	CUT NO. 1	CUT NO. 2

DO	BONDS
	    EST. 1500 M 5.60 $/M
	    FINAL 1410 M = −90 M
		10,100	9,600	− 500	

8	CONCRETE 624 CY
	    EST. 35
	    CONCRETE INC. 33
		21,840	20,590	−1200	

9	REINF. STEEL 1500 CWT
	    EST. 25.00
	    JONES 26.00
		37,500	39,000	+1500	

54	PRECAST GIR'S. 10 EA
	    UNITED 2600
	    MORGAN 2300
		26,000	23,000	−3000
				――――
				− 3200	

Figure 1.4 Cut sheet.

A typical cut sheet is illustrated in Figure 1.4. Those items that require price revisions are listed in column 2, together with their quantities. Space is allowed to enter the subcontractors' quotations below each item. The item numbers are entered in column 1; the bond cost, a direct job overhead item which has no number, is designated as DO.

The cost of each item, as entered on the estimate summary sheet or on the estimate sheets for the items, is listed in column 3. After these costs are entered in column 3, all revisions will be processed on the cut sheet.

A bond cost of $10,100 was entered on the direct job overhead sheet. This cost was based on an estimated total job price of $1,500,000. The rate of change in the bond cost is $5.60 per thousand dollars of total job price, as noted in column 2. The final job price is $1,410,000, and this decrease in price results in the $500 cut shown in column 5.

In estimating the cost of item 8, an estimated price of $35 per cubic yard was used for the concrete cost, since no firm quotation was available at that time. A subsequent quotation of $33 per cubic yard results in the $1200 cut shown in column 5.

Item 9 indicates that an estimated cost of $25 per CWT was used on the estimate summary sheet. A subsequent quotation of $26 per CWT results in an increased cost of $1500, which is listed in column 5.

Item 54 indicates a typical cut resulting from a lower quotation.

Cut number 1, as shown, equals $3200. Later changes in cost are processed using a colored pencil, and the resulting cuts are listed in column 6.

1.10   Equipment Costs

In preparing cost estimates, the cost for heavy equipment is based on an hourly rental rate in the majority of cases. The rental rate used will include the costs of operation, maintenance, insurance, depreciation, and subsistence for the operators when required.

Most contractors do not have a daily or continuing need for heavy equipment, and it is rented when needed from equipment rental firms or from other contractors at prevailing rental rates. Because of competitive pressure in the equipment rental business, the established rental rates are reasonably close to the true cost of owning, operating, and maintaining the particular piece of equipment.

Earthwork contractors frequently own a basic string of earth-moving equipment. In such cases, estimates involving considerable use of the equipment may be made using rental rates based on the contractor's actual hourly costs for owning, operating, and maintaining the equipment. The rates are determined from past job cost records, and in those cases where the contractor conducts an efficient operation, they will be less than the prevailing rental rates.

Prevailing rental rates are always used in estimating job costs when contractor-owned equipment will only be required for short period of time. In such cases the contractor's equipment may be working on another job when needed, or good operators may not be available on a short term basis.

When a contractor rents the heavy equipment as needed, less capital is required and he is not faced with the need to keep the equipment working throughout the year. In the estimates that follow, equipment costs are based on assumed prevailing rental rates.

1.11   Estimating Time Required For Completion

At the time a job is being bid, it is seldom necessary to prepare a formal critical path diagram in order to estimate the time required to complete the project. Those job items on the critical path for the project are normally apparent to the estimator by the time the cost estimates for each item have been completed, and only those items on the critical path are considered. After the bid opening, the successful contractor prepares a time schedule for the job based on a critical path diagram.

The estimate of the time required to complete each job item is made on the estimate sheet for the item after the cost to perform the work has been determined. The simple procedures required to prepare time estimates are illustrated in detail in each of the estimates that follow.

When labor costs represent the major cost element for a given operation, the time required to complete the operation is determined by dividing the total labor cost for the operation by the daily cost of the proposed crew. When equipment costs represent the major cost element, the time required is found by dividing the total number of units to be handled by the daily production rate of the proposed string of equipment.

The working days so determined for each operation are calculated and listed in the "days" column of the estimate sheet, and if necessary some of these times may be reduced by increasing the number or the size of the proposed crews or strings of equipment. The sum of the listed times represents the number of working days required if each operation were performed consecutively. The listed times are then modified to conform to the estimated job scheduling which will entail doing as much of the work concurrently as is practicable. The sum of these adjusted times represents the number of working days required to complete the item.

In a similar manner, the number of working days required for each item on the critical path for the job is tabulated together with an estimated time for mobilization and moving off on completion of the work. Included on this list are the estimated times required to complete certain subcontract items; in some cases conversation with one or more subcontractors may be required to determine a realistic time for their work. The sum of the times listed for the various items represents the number of working days that would be required to complete the job if each item were done consecutively. As before, the times required for each item based on overlapping the work as much as may be practical are listed in a second column and their sum represents the number of working days required to complete the project.

Additional working days are added to allow for the time that will be lost in winter because of inclement weather. The size of this allowance will depend on the geographical area involved; for example, about 20 working days are allowed, per winter, for work performed in Southern California. Assuming 6 holidays per year, the total number of working days is divided by 21.1 to convert the time required for completion to calendar months.

Holidays affect the calendar time required for completion, but do not affect unit costs or production rates. Bad weather increases both unit costs and the time required for completion. The effect of weather conditions on the unit costs used will depend on the circumstances and on the estimator's judgment. Unit costs and production rates from past jobs reflect the effects of the weather on the jobs considered. Disastrous results have been realized, however, when unit costs based on California conditions were indiscriminately used to bid work in Alaska.

The estimated time for completion determines an important component of the total cost for any project since direct job overhead costs depend upon the duration of the job. The time required for completion can be quickly determined when the estimate sheets for the various job items have been properly set up, but the estimator must exercise care and judgment in the process, and must not overlook the importance of the time estimate to the determination of total job cost.

1.12   The Cost System

A cost system based on account numbers for each job item is essential if unit costs are to be developed and if the work in progress is to be properly monitored and controlled. To prevent serious errors of omission, the cost system should be an integral part of the contractor's book-keeping system. It should furnish the contractor the information he requires, and should be standardized as much as possible.

Contractors are inherently highly independent individuals. Their cost-keeping systems are essential to their survival and, in most cases, have been developed by them over the years to fill their specific needs. In many cases, the contractor's cost system and his unit costs are closely guarded. Beginning contractors and young engineers starting a file of unit costs based on their own job experience will, however, need a cost system, and the one described below will be more than adequate.

A list of account numbers is kept at the home office and at each job, and the time for each workman and for each piece of equipment is distributed to the proper accounts in the field on time cards. The distribution may be made by the foremen, using one time card for each crew of workmen. The time cards are extended to dollar amounts in the office and are posted in the cost ledger for the job. All delivery tickets for materials and rented equipment are approved at the job level, and the proper account numbers are written on them at that time.

An account numbering system using two digits to denote each element of the account offers great flexibility and can be adapted to all situations.

Examples of account numbers that might be used are given below, and the time card illustrated in Figure 1.5 is made out using these numbers as an example. In all cases the last two digits of the account numbers will be 01, 02, 03, or 04, and will denote that the cost involved is for labor, material, rental, or subcontract expense.

The account number 64264801 is assumed to read as follows:

	64	job number
	26	rubber gasketed concrete pipe
	48	48 inches in diameter
	01	labor
or	02	material
or	03	rental
or	04	subcontract

The account number 64090001 is assumed to read as follows:

	64	job number
	09	structure excavation
	00	an open sub-account if needed
	01	labor
or	02	material
or	03	rental
or	04	equipment

In the examples above 26 and 09 are account numbers and 48 and 00 represent sub-accounts. Sub-accounts may be further broken down when necessary. Thus the account number 640705030101 is assumed to read as follows:

64	job number
07	bridge concrete
05	bridge number 5
03	abutments and wingwalls
01	set and strip
01	labor

In the example above the list of account numbers for sub-account 03 might be shown as follows:

03	abutments and wingwalls
01	set and strip

ABC Company

Date _____

JOB NO. 64

Account no.	Sub acc't. no.	Description		A. Jones	B. Smith	C. Woods	D. Brown	E. Green	A. Wilson	B. Thomas	ABC #8	Ajax Crane	01 Labor	03 Rental
			Emp. No.								60			
			Class.	L	L	C	C	C	C	E	25			
			Rate ST/OT	9.00	9.00	12.75	12.00	12.00	12.00	13.00			Dollars	
26	48	Conc. pipe – lay		4							4		72.00	240.00
09		Str. ex.									8			200.00
0705	0301	SS wingwalls		4	8	8	8	8					462.00	
01	0801	Backhoe							8				104.00	
			Totals	8	8	8	8	8	8		8 4		638.00	440.00

Figure 1.5 Time card.

07	fine grade
08	pour

The 01, 07, and 08 numbers would be standard for all concrete items and would be quickly understood and memorized by all of the personnel involved.

The account number 6401080101 is assumed to read as follows:

64	job number
01	contractor owned equipment
08	equipment number 8
01	operating expenses
01	labor

It should be noted that B. Thomas, the operator of the company-owned backhoe (ABC #8), is charged to the equipment account, but that the rental cost for the equipment operated and maintained is charged to the structure excavation account. This enables job unit costs to be developed for the equipment indexed to the rental rate charged for the backhoe, and enables the cost of owning and operating the equipment to be developed in the equipment account. In addition, by crediting the equipment account for the rental charged to the jobs, the equipment account will indicate whether or not the contractor is making money on the equipment.

2

Unit Costs

The total costs for the labor and equipment required for each job item are estimated by applying unit costs for labor and equipment to the quantities of those primary activities needed to complete the item. The unit costs are directly proportional to the hourly wage and equipment rental rates. The wage rates used include the costs of all fringe benefits, and the rental rates used include the costs of operation and maintenance together with the costs of subsistence payments to the operators when required.

The costs for labor must be separated from the costs for materials and rentals on the estimate sheets. This is mandatory because certain payroll expenses, discussed in Chapter 3, must be applied to the total labor cost of each job item as a percentage of that labor cost. It is therefore necessary to use a unit cost for labor and a separate unit cost for equipment, when required.

2.1 Unit Costs and Production Rates

Production may be expressed directly in terms of production rates, or indirectly in terms of unit costs. Numerically, the two terms are inversely proportional to one another, but, in the broadest sense, they are interchangeable, and either is easily derived from the other. The unit costs or the production rates for a given primary activity will exhibit wide variation, depending on the type of work being done.

For example, the unit cost to fine grade an abutment foundation will be about 3.5 times the unit cost to fine grade a grade slab, and these costs reflect the relative magnitudes of the quantities normally encountered, the typical subgrade conditions prior to commencing the fine grading operation, and the sizes of the crews normally used.

Similarly, a dragline will perform mass excavation at 3 to 3.5 times the rate achieved when excavating for bridge abutments, and the lower rate reflects the smaller quantities involved and the constant care required to maintain lines and grades.

Therefore, unit costs or production rates must be kept for primary activities relating to the various types of work encountered, and such costs or rates will, in a general way, reflect the typical conditions relating to the particular operation. In all cases, production rates must be indexed to a particular crew size or to a particular piece of equipment, and unit costs must be indexed to the applicable wage or rental rates.

The relationship between unit costs and production rates may be expressed as follows:

$$R = \frac{C}{U} \tag{2.1}$$

or

$$U = \frac{C}{R} \tag{2.1A}$$

where

R = production rate, per hour or per day.
U = unit cost.
C = cost of the crew or equipment per hour or per day.

Equation 2.1 shows that to convert a unit cost into a production rate, a crew or equipment size must be assumed, and that higher production rates will be achieved when larger crews or equipment are used. This relationship is only true within certain practical limits.

To obtain a job, the contractor must bid it at the lowest possible cost, and this requires the balancing of two conflicting requirements. On the one hand, the direct job overhead costs, discussed in Chapter 3, will decrease as the time required for completion is reduced. This aspect, considered separately, seems to indicate the use of large crews. On the other hand, minimum unit costs can only be achieved when the crew and equipment sizes are in proper relationship to the quantities of work to be done.

The job superintendent is responsible for doing the work at the lowest possible cost, and he determines the sizes of the crews and equipment actually used. The unit costs and production rates developed on past work will, in general, reflect optimum crew and equipment sizes. For lowest unit costs, small crews and equipment are used when small quantities are involved, since costs decrease when daily repetitive operations can be established. The proper selection of crew and equipment sizes is critically important, and the ability to make this selection can only be acquired through on-the-job experience.

Production rates are generally used in connection with simple job activities, and must be used when job conditions, which directly affect the resulting unit costs, vary greatly from job to job or from one part of a job to another.

Mass scraper excavation is a simple activity, even though the scraper excavation may be a major item; no secondary activities are involved, the only variable is the round trip cycle time for the scrapers, although the haul distances and the resulting cycle times are subject to wide variation.

Unit costs are generally used in connection with complex activities where the realized unit costs are not greatly affected by the varying conditions found on different jobs.

Since construction cost estimates are expressed in dollars, and since time-of-completion estimates are based, in general, on the costs to do certain portions of the work, unit costs are always used on the estimate sheets. Production rates, when used, are converted into unit costs on a work sheet kept immediately behind the estimate sheet in question.

Unit costs, based on past cost records, will reflect the costs for all incidental overtime, the additional costs for foremen and for different classifications of workmen, and the relative magnitudes of the quantities involved. Overtime costs that can be foreseen, such as the costs to cure concrete over weekends, are estimated separately as a part of the item requiring the overtime.

Job payrolls, cost records, and bank statements are easily reconciled, and since the quantities have been taken off, direct unit costs are readily determined for work under construction.

By comparison, the determination of accurate production rates from job cost records is extremely cumbersome when complex activities are involved. Separate records of the accumulated hours for each classification of workmen must be kept for each primary activity, and these hours must be converted into dollars for purposes of reconciliation.

2.2   *Accuracy of the Cost Data*

The contractor's survival depends, literally, on the accuracy of his cost data. One of management's prime responsibilities is to ensure that cost records for work in progress are kept accurately. On a well organized job, all costs are charged to the accounts to which they belong, and the chips fall where they may. It is impossible to take effective corrective measures if the contractor is unaware of where the trouble lies.

Construction firms have experienced difficulties when dishonest superintendents have falsified cost data. Typically, such superintendents tend to charge costs, properly belonging to an overrunning item for which they are solely responsible, to other large job items where some of the responsibilities are shared by management or by a joint venture partner. When an untested superintendent has an item overrunning the estimate, management must investigate to determine if the trouble is really where the cost records show it to be.

2.3 Overall Unit Costs

Overall unit costs are obtained by dividing the total cost for an entire job item by the number of units in the item. Overall unit costs are used to prepare preliminary estimates and to furnish a rough check when detailed cost estimates are made.

Overall unit costs are inherently imprecise, since the unit costs for the constituent parts of an item and the relative size of these parts will vary. In the preparation of detailed construction cost estimates, overall unit costs may be used in connection with minor or insignificant items.

2.4 Unit Costs for Primary Activities

For maximum precision, unit costs are applied to those primary activities which must be performed to complete each constituent part of the item. The unit cost for a particular primary activity will include the costs for those related secondary activities which must normally be performed to complete the operation. The primary activities are selected so that the relative amounts of the related secondary activities are reasonably uniform from job to job. Combining the costs of the secondary and primary activities greatly reduces the number of activities for which cost records must be kept, simplifies the take-offs and the estimate sheets, and reduces the likelihood of error.

From the general contractor's viewpoint, the primary and secondary activities relating to the construction of reinforced concrete bridge abutments are:

1. Fine grade.
2. Set and strip foundation forms.
   a. Oil forms.
   b. Set and strip starter walls, solid keyways, and templates for reinforcing steel.
   c. Place incidental amounts of expansion joint material.
   d. Move forms ahead to the next pour.
3. Pour foundation slab.
   a. Prepare for pour.
   b. Clean up after the pour.
   c. Cure during regular working hours.
4. Make abutment and wingwall form panels.
5. Set and strip abutment and wingwall form panels.
   a. Oil forms.
   b. Set and strip chamfer and pour strips.
   c. Place incidental amounts of expansion joint material.
   d. Move forms ahead to the next pour.
6. Pour abutments and wingwalls.
   a. Prepare for pour.
   b. Clean up after the pour.
   c. Cure during regular working hours.

In addition to the activities listed above, some abutments may require weepholes, utility openings, or the setting of anchor bolts. Such activities are not required on all jobs, their quantities will vary widely, and they are treated, therefore, as primary activities to which separate unit costs are applied.

Making wall form panels is considered a primary activity, since the quantity to be made will depend on the number of re-uses which are possible, and may not be directly related to the total wall areas. Methods of determining form re-uses are illustrated in Chapters 4 and 5.

2.5 Units of Measure

The units of measure used for unit costs must relate directly to the amount of work to be done. Most labor and equipment costs are directly proportional to the weight of the material to be handled. When this weight is proportional to area, unit costs are, for convenience, often based on areas. Thus, the unit costs for setting and stripping formwork are expressed in dollars per square foot, and the unit costs for wall formwork, presented in Chapters 4 and 5, assume ⅝-inch plyform, 2×4 studs, double 2×4 wales, and taper ties at 2 feet on center, both ways.

Similarly, the cost to set and strip bridge falsework depends on the total weight of material handled, and the unit cost for this work increases with the height of the falsework. Unit costs for setting and stripping wood falsework are expressed in dollars per thousand board feet for falsework of a given height, and the unit costs for setting and stripping steel shoring towers are expressed in dollars per hundredweight of material handled, for towers a given number of frames in height.

Such unit costs exhibit reasonable consistency from one job to another. If the unit costs for falsework were based on the square feet of soffit form supported, no valid comparisons would be possible, except in the case of nearly identical structures.

In all cases, the amount of bridge falsework required is based on a quick preliminary design. This procedure automatically accounts for the additional falsework required under skewed bridges, and when roadways must pass through the falsework.

2.6   Accuracy of the Estimate

Although the completed estimate will be mathematically correct, the estimation of construction costs is not, and can never be, an exact science; the final selection of the unit costs used will, in every case, be based on the estimator's experience and judgment.

The unit costs realized on past work can be determined with a fair degree of precision, but it is impossible to predict the exact unit cost that will be realized on a proposed project. The cost to do work is subject to many variables; from week to week, on the same job, the unit costs for identical work, by the same crews, will vary by 5 percent or more. Among the factors affecting unit costs, are:

1. The work methods used.
2. The amount of access available.
3. The quantities of the work, its complexity, and the opportunity to establish repetitive operations.
4. The size of the crews or equipment used in relation to the quantities of work to be done.

5. The availability of materials.
6. The weather and soil conditions.
7. The quality of workmanship required.
8. The competence of the job inspection and supervision.
9. The availability of competent workmen and of the needed equipment.
10. The degree of cost control exercised.
11. Interruptions by strikes.
12. Union rules.
13. Ground water and water control.

Although the experienced estimator is well aware of the inherent imprecision in selecting unit costs, the selection is based on judgment and cost data, not on guesswork. In the broadest sense, all unit costs and production rates are selected by comparison.

The experienced contractor will have extensive cost records showing his unit costs and/or production rates for the primary activities relating to all of his past work, and his livelihood depends on the completeness and the accuracy of these data.

These unit costs will relate to the particular job conditions under which they were developed and to the contractor's cost system. The unit costs, for a particular primary activity, from several different jobs, may show a maximum variation of as much as 100 percent. The reasons for the extreme variation are usually evident, and reflect unusually small quantities of work in some cases, severe job conditions in others, and, sometimes, poor cost data.

When an array of past unit costs for a given primary activity is considered, a probable value for the average unit cost applicable to average conditions can be selected, and the variation for unit costs relating to average conditions will normally range from 5 to 10 percent. The true worth of the experienced estimator lies in his ability to select a reasonable unit cost from the array, to consider the job conditions to which it applies, and to modify it, as necessary, for use on a proposed project with its particular job conditions.

In all cases, extreme care must be exercised when large quantities are involved.

Some individuals in the construction industry decry the use of unit costs. These individuals have one thing in common—they have never personally made a detailed construction cost estimate.

The use of unit costs is the only rational method for estimating construction costs. The procedure must be based on skill, experience, cost data, and judgment—it can never be precise, but use is made of all available data and information. The thousands of successful contractors are solid evidence that the procedure works.

2.7   Warning Regarding Unit Costs

Each contractor's unit costs and production rates are unique, and all unit costs and production rates are dangerous if indiscriminately used.

In the following chapters, average unit costs and production rates are presented for many types of work performed by general contractors. These costs assume average job conditions for work done in a temperate climate under the continuous supervision of a good superintendent. These unit costs are shown to illustrate the procedures followed in estimating construction costs.

It is virtually impossible to list every secondary activity whose cost is included, or to enumerate all the typical job conditions under which the unit costs were realized. The use of the unit costs presented is illustrated, however, in many examples, all of which pertain to typical job situations.

The unit costs presented herein should not be blindly used by inexperienced estimators. When large quantities are involved, the unit costs used should, in every case, be based on the estimator's personal cost records for similar work performed under known job conditions. When moderate quantities are involved, the unit costs presented herein should only be used by estimators who have the experience and background to confirm their magnitude by consideration of the estimator's own cost data for similar work.

2.8   Indexing Unit Costs

Unit costs developed on past jobs having various wage and equipment rental rates must be converted to be applicable to a proposed project having different rates.

The conversion process is simple.

A series of indexes is assigned to all jobs, past or proposed and, as illustrated in the estimates in Chapter 4, the indexes used are written at the top of the estimate sheets for each job item. The labor index is the cost of an assumed crew per hour or per day, and the equipment rental indexes are the hourly rental rates for the equipment. The unit costs are directly proportional to the applicable indexes.

In the following chapters, average unit costs and their related indexes are given for many types of work performed by general contractors. In

Chapters 4 and 5, for example, the indexes are based on the following wage and rental rates, and the wage rates include all fringe benefits.

Carpenter	$16.34 per hour
Laborer	$13.48 per hour
Cement finisher	$16.10 per hour
35 ton truck crane	$75.00 per hour

The labor index used for setting and stripping formwork is the cost of a crew of six carpenters plus two laborers, working for 8 hours, and is designated K_{6+2}. Based on the above rates, $K_{6+2} = 1000$. For a proposed job where $K_{6+2} = 1100$, those unit costs based on $K_{6+2} = 1000$ would be multiplied by $\frac{1100}{1000} = 1.10$. Similarly, unit costs from a past job indexed to $K_{6+2} = 700$ would be multiplied by $\frac{1100}{700} = 1.57$.

Where, with the passing years, the wage rates for carpenters and laborers increase at approximately equal percentage rates, the K_{6+2} index will convert unit costs with no error in any case. Where one of the two wage rates increases at an appreciably faster rate than the other, a small error will be introduced into the converted unit costs over a period of years in those cases where the actual composition of the crews used is appreciably different from the 6 carpenters plus 2 laborers assumed. This error should not exceed 2 percent over a period of 8 or 10 years and in all likelihood would be much less. In practice unit prices are being determined and updated continuously and the contractor will give more weight to his more recent cost records. In such cases the error from this source will be negligible.

Where work is primarily done with laborers the K_L index which is the hourly wage rate for 1 laborer is used. The unit costs presented herein for such work are indexed to $K_L = 13.48$.

Where work is done with laborers and cement finishers in approximately equal numbers the K_{L+F} index is used, which is the hourly wage rate of 1 laborer plus 1 cement finisher. The unit costs for such work presented herein are indexed to $K_{L+F} = 29.58$.

The wage rate for a given trade may vary with the type or classification of the work being performed. Thus a vibrator operator may receive more pay while vibrating than when he is performing as a general construction laborer. In such cases, the wage rate used in indexing unit costs is the prevailing rate for the majority of the workmen involved, and the premium paid to those engaged in special activities will be automatically included in the unit costs developed.

The unit costs for concrete-placing equipment, presented in Chapters 4 and 5, are indexed to $K_{Cr} = 75$, which is the hourly rental rate for a 35-ton truck crane, operated and maintained. The equipment costs for a number of miscellaneous activities are indexed to K_{BH}, which is the hourly rental rate for a small rubber-tired loader-backhoe. The costs to place concrete with concrete pumps are discussed in section 5.1.4.3.

2.8.1 On-And-Off Charges for Equipment

The unit cost for doing work with equipment depends upon the hourly rental rate charged, the production rate achieved, and the costs to move the equipment on and off of the job. Ideally, on-and-off charges should be estimated separately, and this is done whenever the number of on-and-off moves can be predicted with reasonable accuracy. Thus the costs to move earth-moving equipment on-and-off the job are always considered as a separate item of expense.

On the other hand, when equipment is required to place concrete, the number of on-and-off moves required is often difficult to predict, and the unit costs for this work, presented in Chapters 4 and 5, include 1 hour of travel time. As noted on the unit cost tables, these unit costs assume pours of average size for the type of work being considered.

2.8.2 Increase in Unit Costs With the Passage of Time

Increases in unit costs due to rising wage and rental rates are easily determined, as outlined above. However, unit costs for a given activity will also slowly increase with the passage of time, because of the decreasing productivity of the workmen. Such cost increases tend to be slow and irregular, cannot be predicted with much accuracy, and in most cases are within the limits of accuracy of the unit cost selection process. In practice, the estimator places more weight on his more recent unit costs, and this procedure accounts for the decreasing productivity of the crews.

2.9 Converting Unit Costs to Metric Units

English and metric units are readily converted one to the other. When making such conversions, it is essential to differentiate between sentences and equations. Thus, "1 inch = 2.54 centimeters" is a correct statement or sentence. The corresponding equation, however, is

$$\text{inches} \times 2.54 = \text{centimeters}.$$

To avoid confusion, the following conversion factors are expressed as equations:

Inches	$\times$ 2.54001 = centimeters
Feet	$\times$ 0.304801 = meters
Yards	$\times$ 0.914402 = meters
Square inches	$\times$ 6.45165 = square centimeters
Square feet	$\times$ 0.0929034 = square meters
Square yards	$\times$ 0.836127 = square meters
Cubic inches	$\times$ 16.3873 = cubic centimeters
Cubic feet	$\times$ 0.0283172 = cubic meters
Cubic yards	$\times$ 0.764559 = cubic meters
MFBM	$\times$ 2.35975 = cubic meters
Pounds	$\times$ 0.453592 = kilograms
Tons (2,000 pounds)	$\times$ 907.185 = kilograms
Centimeters	$\times$ 0.393699 = inches
Meters	$\times$ 3.28083 = feet
Meters	$\times$ 1.09361 = yards
Square centimeters	$\times$ 0.155000 = square inches
Square meters	$\times$ 10.7639 = square feet
Square meters	$\times$ 1.19599 = square yards
Cubic centimeters	$\times$ 0.0610229 = cubic inches
Cubic meters	$\times$ 35.3142 = cubic feet
Cubic meters	$\times$ 1.30794 = cubic yards
Cubic meters	$\times$ 0.423774 = MFBM
Kilograms	$\times$ 2.20462 = pounds
Kilograms	$\times$ 0.00110231 = tons (2000 pounds)

Unit costs expressed in dollars and English units are converted to unit costs expressed in dollars and metric units as follows:

Multiply unit costs in dollars per	by	To obtain unit costs in dollars per
Foot	3.281	meter
Square foot	10.76	square meter
Square yard	1.196	square meter
Cubic foot	35.31	cubic meter
Cubic yard	1.308	cubic meter
Pound	2.205	kilogram
Ton (2,000 pounds)	0.001102	kilogram
MFBM	0.4238	cubic meter

2.10 Converting Unit Costs to Other Currencies

Unit costs expressed in dollars are easily converted to unit costs expressed in another currency if the wage rates, expressed in the other currency, are known:

$$C' = C_d \times \frac{K'}{K} \qquad (2.1)$$

where:

C' = unit cost expressed in a foreign currency.
C_d = unit cost expressed in dollars.
K' = the appropriate unit cost index based on the foreign currency and wage rates.
K = the appropriate unit cost index based on dollars.

For example, assume that a unit cost of $0.20 per square foot, indexed to $K_{6+2} = 1000$, is to be converted to a unit cost expressed in francs per square meter, and that the converted unit cost will be used in connection with work where the wages are paid in francs. Assuming the wages in francs to be

> Carpenters: 27.25 francs per hour
> Laborers: 22.50 francs per hour

then $K'_{6+2} = 27.25 \times 6 \times 8 + 22.50 \times 2 \times 8 = 1668$, and, from equation 2.1, we have

$$c' = (0.20 \times 10.76) \times \frac{1668}{1000} = 3.59 \text{ francs per square meter}$$

In this equation, the factor 10.76 is taken from the preceding tabulation, and it converts the unit cost expressed in dollars per square foot to one expressed in dollars per square meter.

When, for estimating purposes, unit costs are converted to other currencies, the conversion depends solely on the factors set forth in equation 2.1. The actual monetary exchange rates for the currencies involved are not relevant, since the wage and rental rates paid in different countries will be different, even when these rates are expressed in terms of a common currency.

It is essential to realize that unit costs determined from cost records for work done in the United States are based on using American crews and methods, while working with American specifications under particular climatic conditions. The estimator must therefore apply judgment to converted unit costs in order to correct for differences in the anticipated efficiencies of the crews, methods specified, and climate.

3

General Job Costs

General job costs are those costs incurred from running the job as a whole. These costs are not directly related to the performance of any particular item of work. They consist of direct job overhead, payroll expenses, the job's share of the contractor's general overhead, and the job profit.

3.1 Direct Job Overhead

Direct job overhead is the direct cost of running the job. Before completing the direct job overhead sheet, the estimator must have an estimate of the time required to complete the work, a close approximation of the total job cost, and complete familiarity with all of the job items. For these reasons the direct job overhead sheet is the last sheet set up by the estimator.

3.1.1 Check List for Direct Job Overhead

Direct job overhead includes the costs of those items on the following list that apply to the particular job being bid.

1. Contract bonds.
2. Insurance.
3. Permits, fees, and tests.

4. Supervision and job office salaries.
5. Clearing and grubbing.
6. Office and sheds.
7. Power.
8. Water.
9. Telephone.
10. Radios.
11. Sanitary facilities.
12. Drinking water.
13. Vehicles.
14. Hoisting.
15. Job signs.
16. Layout.
17. Watchmen.
18. Access roads.
19. Fences.
20. Heating.
21. Medical and safety.
22. Punch list.
23. Clean up.
24. Traffic control.
25. Wage raises.
26. Subsistence.
27. Special requirements.

3.1.1.1 CONTRACT BONDS. Contract bonds were discussed in Section 1.4.4. In every case, the costs for the bonds are obtained from the contractor's bonding company, since both the rates and the work classifications to which they apply are subject to change.

On lump sum jobs, as illustrated in Figure 1.1, the bond cost is usually the second item on the estimate summary sheet. On unit price jobs, the bond cost is usually the first item on the direct job overhead sheet, as illustrated in Figure 4.7.

At the time the estimate sheets are set up the cost for the bonds is based on an approximate job price, and the bond cost is revised, if necessary, when the final job price is determined. The bond costs shown on the estimates presented in Chapter 4 are based on an assumed rate of $7.50 per thousand dollars of total job price.

3.1.1.2 INSURANCE. The costs for insurance represent a significant portion of the overall job cost. For bookkeeping purposes, all insurance costs, with the exception of the cost for the contractor's equipment floater insurance, are considered as direct job overhead items.

The costs for workmen's compensation insurance, public liability and property damage insurance and, in some cases, the costs for all liability insurances, are based on the contractor's payroll. For estimating purposes, the costs for these payroll insurances are determined and entered on the estimate sheets for each job item as explained in the following section on payroll expenses.

Those insurance costs not based on the payroll are carried on the estimate summary sheet, on building work, and on the direct job overhead sheet on unit price work. In both cases the charges entered initially are based on estimated job costs, and are revised after the final job cost is determined.

Workmen's Compensation Insurance. The cost for this insurance is always based on the job payroll, as discussed in the section on payroll expenses.

Public Liability and Property Damage Insurance (Exclusive of Automobile Insurance). The costs for this insurance are also based on the job payroll, as discussed in the section on payroll expenses.

Contingent Liability and Property Damage Insurance. This insurance indemnifies the contractor in the event of claims for injuries or damages resulting from the activities of subcontractors or other parties, for which the contractor may be held responsible. The cost for this insurance may be expressed as a percentage of the total price of the subcontracted work, or it may be included in a composite rate for all liability insurances, which is applied to the job payroll.

When the cost is based on a percentage of the total price of subcontract work, the percentage rate used is normally the same for all subcontracts on a given job.

Products and Completed Operations Liability Insurance. This insurance indemnifies the contractor in the event of claims for injury or damages to third parties arising from the normal use of the completed project. The cost for this insurance may be expressed as a percentage of the total job price, or it may be included in a composite rate for all liability insurances, which is applied to the job payroll.

Public Liability and Property Damage Insurance on Automobiles and Pick-up Trucks. The cost for this insurance may be a stated amount per vehicle, per year, or it may be included in a composite rate for all liability insurances, which is applied to the job payroll.

When the cost is on a per vehicle basis, the cost may be included in the rental rate charged for the vehicle on the direct job overhead sheet.

Excess Limits Insurance or Umbrella Insurance. Either of these insurances provide increased limits for the primary liability insurances. An umbrella policy will, in addition, provide certain additional coverages which are usually subject to a deductible of $10,000 or more. The costs for the umbrella policy have become excessive, and the excess limits insurance is often used. The costs for these insurances are a percentage of the total premium charged for all of the primary liability insurances, subject to a minimum premium, but the cost may be included in a composite rate used for all liability insurances, which is assessed to the job payroll.

Fire Insurance. Fire insurance is always required on building work, although its cost is sometimes paid by the owner. The cost will be a specified percentage of the bid price for that portion of the work which is subject to fire damage.

Installation Floater Insurance. This insurance indemnifies the contractor in the event of job accidents resulting in damages to portions of the completed work. The insurance may be purchased for protection against a particular hazard, or it may cover all risks. The price is quoted by the insurance agent for each specific job.

The cost for installation floater insurance is often prohibitive. In such cases, the cost for those additional steps necessary to substantially reduce the risks from particular hazards is included in the costs for the job item involved.

Contractor's Equipment Floater Insurance. This insurance indemnifies the contractor should his equipment be damaged by accident or vandalism. In most cases, only motorized equipment is insured, and the insurance costs are included in the rental rates charged for the equipment.

3.1.1.3 CLEARING AND GRUBBING. The cost of incidental clearing and grubbing is included here in those cases where no job item was set up for this work.

3.1.1.4 OFFICES AND SHEDS. This item covers the costs of setting up and removing the job office, sheds, and yard.

3.1.1.5 POWER. This item covers monthly power bills. The costs to connect and disconnect the power source are included in the cost of the job item requiring power.

3.1.1.6   WATER.   This item covers the monthly water bills for incidental amounts of water. The costs of developing the water supply and applying such water are included in the cost of the item requiring the water. The costs relating to large amounts of water like those required for earthwork compaction are included in the cost of the job item involved or in special job items for these costs.

3.1.1.7   RADIOS.   This item covers the monthly rental charges for the mobile radios used.

3.1.1.8   SANITARY FACILITIES.   This item covers the cost of the job toilets and other sanitary facilities as required. The cost is based on monthly rental rates which include servicing and moving the facilities when necessary.

3.1.1.9   DRINKING WATER.   This item covers the cost of water cans, ice, and paper cups used to supply drinking water to the crews.

3.1.1.10   VEHICLES.   This item covers the monthly rental charged for pickup trucks, flat rack trucks, and the superintendent's car. The rates used include fuel, maintenance, repairs, and depreciation.

3.1.1.11   HOISTING.   This item covers the costs of hoisting men and materials. It is primarily of importance on building work, where the specifications frequently require the general contractor to furnish hoisting facilities for all of the trades.

On buildings up to three stories in height, personnel elevators are not required and materials hoisting may be done with truck cranes. Although the general contractor will have included the costs for hoisting in his cost estimates for his own work, the requirements of each subcontractor must be considered and an allowance made for the additional crane rental involved.

On tall buildings one or more hoisting elevators are required for the workmen and for materials. On larger work tower cranes may be used to handle materials. In both cases the equipment is usually set up initially by the supplier for a fixed fee and rented to the general contractor on a monthly basis.

In all cases the operators are paid by the contractor who also adds to the height of the elevator and jumps the tower crane up with his own forces as the building rises. On completion of the work the contractor usually takes down the equipment and returns it to the supplier, using his own crews for the dismantling operation.

3.1.1.12 LAYOUT. This item covers the cost of job layout when it is not furnished by the owner. Incidental survey work by job personnel is not considered.

3.1.1.13 FENCES. This item covers the cost of temporary fences required to protect the job site. Permanent fencing is carried as a separate job item and is normally furnished and installed by a subcontractor.

3.1.1.14 HEATING. This item covers the cost of equipment and materials required to protect the work from cold weather damage.

3.1.1.15 PUNCH LIST. The punch list is the list of minor repairs and corrections that must be completed prior to final acceptance of the work by the owner. This item is only significant on building work and will depend on the quality of the job supervision and inspection. On office buildings and schools, where $K_{6+2} = 1000$, the cost may run about $35 per MSF of floor area for the general contractor's own work. Each subcontractor completes his own punch list items at his own expense.

3.1.1.16 CLEANUP ON ENGINEERING WORK. On engineering work this item covers the cost of the final cleanup required to obtain acceptance of the work by the owner. The work consists primarily of dressing up the job area with a blade, and the costs of such work are discussed in the chapters on earthwork. No additional charges for day-to-day cleanup are required on engineering work, since these costs are included in the unit costs developed for the various items of work.

3.1.1.17 CLEANUP ON BUILDING WORK. The cost of cleanup is a significant item on building work; it includes the cost of day-to-day general cleanup in addition to the cost of the final cleanup required by the specifications.

On building work there are many subcontractors and the work is done indoors. Even though each subcontractor is required to clean up his own work on a continuing basis, there is an inevitable accumulation of dirt and debris that must be removed by the general contractor in the interests of good housekeeping and overall job efficiency.

The final cleanup required on building work is always covered in the specifications and normally includes the following items:

1. Dressing up the exterior site. This work is similar to that required on engineering work.
2. Sweeping all floors, dusting cabinets, doors and electrical fixtures and cleaning toilet fixtures.
3. Washing windows and glass doors and waxing and polishing some floors. Such work is normally done on a subcontract basis by a janitorial firm.

In view of the variables involved, it is not possible to estimate the cost of cleanup on building work with much precision. It is probable that the costs actually realized by the same contractor on two similar building jobs having different superintendents and subcontractors would vary by as much as 20 percent. As an order of magnitude, the labor cost of the day-to-day cleanup plus the sweeping, dusting, and cleaning noted in 2 above has run about $190 per MSF of floor area on buildings ranging from 80 MSF to 200 MSF in size, where $K_L = 13.48$.

The proposed cost for cleanup labor should be considered with respect to both the floor area, and the number of laborer hours represented per day of job duration. The final cost selected will be based on the estimator's judgment, augmented by cost records from similar work.

Since burning is now prohibited in most areas, an additional allowance must be made for removing the debris from the job site. In most cities there are firms that supply debris boxes and dispose of their contents as required on a monthly rental basis.

3.1.1.18   TRAFFIC CONTROL.   This item covers the cost of the flares, barricades, traffic control signs, and flagmen needed to protect the public during the course of the job. The flares, barricades, and signs are normally rented, and the rental charge includes maintenance and replacement as required. When possible, traffic control costs should be included with the cost of the primary job item requiring such control.

3.1.1.19   WAGE RAISES.   This item covers the additional costs for labor and equipment resulting from wage raises that occur during the course of the job. In such cases, a portion of the work will be done under the existing rates, and the remainder will be done under the new higher rates. Wage raises are listed on the direct job overhead checklist as a reminder. The additional costs resulting from wage raises are estimated on the estimate sheets for each job item.

The date on which the wage raise will occur must be located on the time schedule for the job. As discussed in Section 1.11, the estimated time required for completion and the job time schedule can only be determined after the cost estimates for each job item have been completed.

When the wage raise occurs during the later stages of the job, the costs for all of the job items are estimated on the basis of the existing rates, a job time schedule is then set up, and the point on the schedule where the wage raise will occur is determined.

The labor and equipment costs for each job item or portion of a job item that will be performed after the wage raise are then increased by an appropriate percentage. The percentage used is determined from the ratio of the cost indexes (Section 2.8) applicable after and before the wage raise. Thus if $K_{6+2} = 1000$, based on the existing rates, and if $K_{6+2} = 1075$ after the wage raise, then those labor costs for carpentry work occurring after the raise would be increased by 7.5 percent.

When most of the job is done after the wage raise is in effect, the estimated cost of all of the job items is based on the increased rates, and deductions are made on the estimate sheets for the work that will be performed under the existing rates.

The effect of wage raises on rental rates depends on decisions made by the owner of the equipment. A raise of $1.50 per hour for an equipment operator will often result in an increase of $3.00 per hour in the rental rate charged.

3.1.1.20 SUBSISTENCE. Some contractors convert the specified daily payments for subsistence, when required, into hourly rates that are treated as additional wages. In all cases, the applicable hourly rates are added to the rental rates charged for operated equipment.

The requirements for paying subsistence are somewhat complex, and it is not possible to convert a specified daily rate into an exact hourly rate; therefore, the following procedure is recommended for estimating the costs of subsistence payments for job labor.

The total labor cost (exclusive of payroll expenses, Section 3.2) for each job item is increased by an appropriate percentage to cover the estimated costs for subsistence. The composition of a typical crew for each job item is assumed, and the percentage used is based on the ratio of the estimated daily costs for the crew with and without subsistence. The allowances made for subsistence on labor are then collected and monitored as a direct job overhead item, for control purposes, on those jobs awarded to the contractor.

This procedure results in unit costs for labor that are not affected by variations in subsistence payments, and that can be interpreted on their own merits. Similarly, the allowance made for subsistence can be monitored on its merits, and adjusted on future estimates, if warranted.

3.1.1.21 SPECIAL REQUIREMENTS. The preceding list contains those direct job overhead items that occur regularly in practice. This item includes the costs of additional direct job overhead items that may be required by a particular set of plans and specifications.

3.2 Payroll Expenses

Payroll expenses are the costs of certain taxes and insurances which are based on the contractor's total payroll costs for all jobs in progress. These expenses are allocated to each job in proportion to its payrolls, and are monitored against the allowances made on the estimate, for control purposes. The cost of the payroll expenses is estimated by applying a single composite percentage rate to the total labor cost for each job item.

The true cost of each payroll expense item is determined by applying a specified percentage rate to a dollar amount which is based on the employee's base wage rate plus the hourly rate for the employee's vacation plan. For purposes of this discussion, dollar amounts so determined are defined as "earnings."

From the contractor's point of view, however, the wage rate for an employee working under a typical union contract includes the base wage rate plus the hourly rates for all fringe benefits. As discussed in Chapter 2, all unit costs for labor and the estimated costs for all job items are based on wage rates so defined.

The hourly rates for each fringe benefit are multiplied by the hours worked, and these amounts are paid into trust funds set up by the union for each fringe benefit. The rates used for fringe benefits are the same regardless of whether the hours worked are at straight time or overtime. Typical fringe benefits are:

1. Vacation plans.
2. Pension plans.
3. Health and welfare plans.
4. Other minor plans such as apprenticeship training or industry promotion.

On nonunion, prevailing-wage contracts, all of the fringe benefits are paid directly to the workmen. Premium pay is normally paid for overtime work, and the premium pay rate is a percentage of the base wage rate. Thus, when time and one half is paid for overtime, the rate of premium pay will be 50 percent of the base wage rate, and when double time is paid, the rate of premium pay will be 100 percent of the base wage rate.

The typical components of the payroll expense item are discussed below, and a method is developed whereby the specified percentage rates for each payroll expense, based on earnings, are converted into a single, composite rate, applicable to wages. This composite rate is then applied to the total estimated labor cost for each job item.

3.2.1 Payroll Taxes

Payroll taxes normally consist of federal social security taxes (Federal Insurance Contributions Act, or F.I.C.A.), federal unemployment taxes, and state unemployment taxes. The specified percentage to be charged for each of these taxes is applied to a dollar amount based on the base wage rate plus the vacation plan rate, plus the premium pay rate, when applicable.

Assuming the following conditions:

Base wage rate: $10.00 per hour
Vacation plan rate: $1.50 per hour
Overtime: Time and one half
Straight time hours: 40
Overtime hours: 6
Total hours worked: 46

the premium pay rate for overtime would be 50 percent × $10.00 = $5.00, and the dollar amount to which the percentages for taxes would be applied would be 46 × ($10.00 + $1.50) + 6 × $5.00 = $559.00.

In the following discussion of the individual payroll taxes, the dollar amount calculated as above is referred to as "earnings including premium pay."

3.2.1.1 FEDERAL SOCIAL SECURITY TAXES (F.I.C.A.). Only that portion of the social security tax which is paid by the employer is considered in determining the allowance for payroll expenses.

Each employee also pays a portion of the social security tax. This portion is withheld by the employer, and is paid to the Internal Revenue Service for credit to the employee's social security account. Except for the overhead involved, the tax paid by the employee does not constitute an additional expense to the contractor.

In our estimates, we will assume that the employer's portion of the social security tax is 5.85 percent of the first $16,500 of earnings, including premium pay, paid to each employee.[1]

---

[1]1979 rates are 6.13 percent of the first $22,900.

3.2.1.2 FEDERAL UNEMPLOYMENT TAX. In our estimates, we will assume this tax to be 3.2 percent of the first $4,200 of earnings,[2] including premium pay, paid to each employee, and will assume further that the 3.2 percent rate may be reduced by 90 percent of the rate used in determining the state unemployment tax, up to a maximum deduction of 2.7 percent.

3.2.1.3 STATE UNEMPLOYMENT TAX. In our estimates we assume this tax to be 3.6 percent of the first $7,000 of earnings, including premium pay, paid to each employee.

3.2.2 Payroll Insurance

The costs for workmen's compensation insurance and for all or part of the contractor's liability insurance are determined by applying specified percentage rates to the employee's earnings.

3.2.2.1 WORKMEN'S OR WORKERS' COMPENSATION INSURANCE. This insurance covers medical and other costs resulting from job-related accidents or afflictions. The cost for this insurance has increased drastically in the past few years, and it now represents the major portion of the payroll expenses.

Workmen's compensation insurance is mandated by the states; their procedures for assessing its cost may vary, and the percentage rates charged are subject to frequent changes. The applicable rates and procedures must, therefore, be obtained from the insurer in all cases. In the following discussion, we assume the job is located in California.

The percentage rates used to determine the costs for this insurance vary greatly, on a given job, depending on the particular classification of work being performed. The actual costs are determined by an audit of the contractor's books, and it is essential that the work classifications for the workmen be noted on the daily time cards.

The percentage rates for workmen's compensation insurance are applied to a dollar amount based on the base wage rate plus the vacation plan rate, and premium pay for overtime work is not included. Thus referring to the example previously given in Section 3.2.1, the dollar amount to which the percentages for workmen's compensation insurance would be applied is

$$46 \times (\$10.00 + \$1.50) = \$529.00.$$

The dollar amount so calculated will be referred to as "earnings excluding premium pay."

---

[2]1979 rates are 3.4 percent of the first $6,000.

In our estimates, we will assume the following percentage rates for workmen's compensation insurance:

Classification	Percentage of earnings excluding premium pay
Concrete work NOC	12.5
Concrete bridges and culverts	28.5
Demolition work	30.0
Carpentry work	15.0
Earthwork and roadwork	6.5

3.2.2.2 LIABILITY INSURANCE. The several types of liability insurance normally carried by contractors were discussed in Section 3.1.1. In our estimates we will assume a composite rate of 2.8 percent for the following insurances:

1. Public liability and property damage (including automobiles).
2. Contingent liability and property damage insurance.
3. Products and completed operations liability insurance.
4. Excess limits insurance.

The 2.8 percent rate is usually applied to earnings, including premium pay, as defined previously in the section on Payroll Taxes. Insurance rates and terms will vary, however, with the insurer used. In all cases, the specifications must be checked for special requirements which might affect the rates or the limits for the insurance.

3.2.3 The Allowance for Payroll Expenses

The contractor's cost estimates for each job item are based on wage rates which include the base wage rate plus all fringe benefits. As discussed above, the specified percentage rates for payroll taxes and insurances are applicable to earnings which are based on the base wage rate plus only the vacation plan. Premium pay for overtime is included in some instances and excluded in computing the cost of workmen's compensation insurance; all earnings in excess of certain stated amounts are excluded in determining the costs for payroll taxes.

The final costs for all of the payroll expense items are determined by audits. For estimating purposes, however, the contractor must determine a single composite percentage rate for all payroll expense items, and this rate must be applicable to the total labor costs shown on the estimate sheets for each job item.

Based on the assumed percentage rates and limits outlined above for payroll taxes and insurances, a reasonable approximation of the composite percentage rate for payroll expenses is easily determined. Although state taxes will be different for the various states and the rates for all taxes and insurances tend to increase with the passage of time, the method for determining the composite rate for payroll expenses, outlined in the following example, will be applicable in all cases.

The assumed hourly earnings and wage rates are shown in Table 3.1, and these rates are used in the cost estimates presented in Chapter 4.

The total yearly earnings shown in Table 3.2 are based on the assumption that the average hourly-paid workman will be in our employ for only 3 months or 520 hours during the year, whereas the superintendent, a salaried employee, will work all year long. This assumption is reasonable for a beginning contractor bidding his first small job. Established contractors doing larger jobs may find that their average workman is employed by them for 8 to 10 months during the year, and this will reduce the taxes expressed as a percentage of earnings, in Table 3.2, by as much as 3 percent.

Established contractors, however, keep cost data showing the total wages paid during each past year, and the total costs for each of the payroll taxes. The percentages, based on wages, paid for payroll taxes in the past year are then determined directly, and are adjusted to fit the rates and limits applicable to the current year.

In Table 3.3 the specified percentage rates for payroll taxes and insurances are converted into rates applicable to the estimated labor costs. The factors used for the conversion are taken from Table 3.1, and are the ratio of the hourly earnings to the hourly wages for each classification of employees. The percentage rate for workmen's compensation insurance will vary, depending on the type of work being done. The rate for concrete work, NOC (not otherwise classified), is used for illustration in Table 3.3. Since the cost for workmen's compensation insurance now constitutes a major portion of the payroll expense item, and since the rates charged for this insurance vary widely for different classifications of work, the total percentage to be allowed for payroll expenses must be calculated for each classification of work that the contractor will perform.

Assuming the composition of a typical crew to be six carpenters plus three laborers plus one cement finisher, the percentage to be allowed for payroll expenses for concrete work, NOC, will be

$$(6 \times 19.5 + 3 \times 18.7 + 1 \times 19.9) \div 10 = 19.3 \text{ percent},$$

and this percentage is applied to the total labor cost shown on the estimate sheets for such work. In view of the assumptions required to determine

Table 3.1. Earnings and Wages in Dollars per Hour

Classification	Base Wage Rate	Fringe Benefits		Health and Welfare		Earnings[a]	Total Wages[b]
		Vacation	Pension		Other		
Carpenter	$11.69	$1.00	$1.95	$1.60	$0.10	$12.69	$16.34
Laborer	9.28	0.70	2.20	1.20	0.10	9.98	13.48
Cement Finisher	11.50	1.25	2.00	1.25	0.10	12.75	16.10
Superintendent	13.00	0.75	—	—	—	13.75	13.75

[a]Earnings = base wage rate + vacation plan rate.
[b]Total wages = base wage rate + all fringe benefits.

63

Table 3.2.   Payroll Taxes Paid

Classification	Carpenter	Laborer	Cement Finisher	Superintendent
Total yearly earnings (in our employ)	$6600	$5200	$6600	$28,600
Type of tax			Taxes Paid	
F.I.C.A.[a]	$ 386	$ 304	$ 386	$ 965
F.U.I.[b]	$ 21	$ 21	$ 21	$ 21
S.U.I.[c]	$ 238	$ 187	$ 238	$ 252
Total taxes	$ 645	$ 512	$ 645	$ 1,238
Taxes as a percentage of yearly earnings	9.8%	9.9%	9.8%	4.3%

[a]5.85% of first $16,500.
[b](3.2–2.7)% of first $4,200.
[c]3.6% of first $7,000.

this percentage rate, it would, in practice, be rounded up to 19.5 percent. It is essential that the actual costs for all of the payroll expense items be monitored, on an on-going basis, against the amounts recovered by the application of the percentage rate used.

In the example above, no allowance was made for the effect of premium pay for overtime, even though the costs for normal overtime are reflected in the unit costs and will appear in estimated costs for the job items. The

Table 3.3.   Payroll Expenses Expressed as a Percentage of the Estimated Labor Costs

Classification	Carpenter	Laborer	Cement Finisher	Superintendent
Taxes	9.8%	9.9%	9.8%	4.3%
Compensation insurance	12.5%	12.5%	12.5%	12.5%
Liability insurance	2.8%	2.8%	2.8%	2.8%
Total percentage based on earnings[a]	25.1%	25.2%	25.1%	19.6%
Factor	$\frac{12.69}{16.34}$	$\frac{9.98}{13.48}$	$\frac{12.75}{16.10}$	1
Total percentage based on wages[b]	19.5%	18.7%	19.9%	19.6%

[a]Earnings = base wage rate + vacation plan rate.
[b]Wages = base wage rate + all fringe benefits.

amount of overtime worked is normally small with respect to the total hours worked, and its effect can be neglected in all typical situations.

3.3 General Overhead and Profit

General overhead is the cost of running the business as a whole, and is not directly related to the performance of any specific job. General overhead includes the cost of the following items:

1. Owner's salary.
2. Home office salaries.
3. Home office rent.
4. Home office power and water.
5. Home office telephone and radios.
6. Home office vehicles.
7. Home office supplies and repairs.
8. Depreciation.
9. Contractor's yard.
10. Professional services.
11. Business taxes and licenses.
12. Insurance.
13. Health insurance for salaried personnel.
14. Dues and subscriptions.
15. Travel and entertainment.
16. Bidding expenses.
17. Payroll expenses on general overhead labor.
18. Donations.
19. Special government programs.

The monthly or yearly cost of the contractor's general overhead is easily estimated, but the amount to be charged to any given job depends upon the total volume of work done during the course of the year.

The contractor has no way to predict precisely the volume of work that will be awarded to him during an upcoming year. This problem is common to all business enterprises but is, perhaps, more severe in the construction industry; the sizes of the jobs bid vary greatly, the number of jobs awarded per year is relatively small, and a difference of one job of average size will often result in a significant variation in the volume of work done during the year.

In practice, the contractor decides on an amount to be added to the bare cost of each job for general overhead and profit, and then sets a goal to obtain sufficient work to pay the general overhead and to make a profit over a year's time.

The bare cost of any job is the total cost of doing the work including direct job overhead, but does not include the allowance for general overhead and profit. The amount allowed for general overhead and profit is the contractor's markup; it may be expressed in dollars or as a percentage of the bare cost of the job.

The amount of the contractor's markup is limited by the competition from fellow contractors, and will decrease as the number of bidders increases.

Preliminary cost estimates prepared by architects and engineers frequently contain an item for contingencies. Except in the case of rare and unusual circumstances, the contractor never includes an allowance for contingencies as a separate item of cost in bidding work. To include such an item would result in a high bid in most cases. Each job item is bid as the estimator sees it, and the costs of all conditions that can reasonably be expected to occur are included in the estimated costs of the items involved.

Those increased costs that would result if unlikely events were to occur are considered by the contractor at the time the size of the markup is determined. Where such costs represent a small percentage of the normal markup, they are usually ignored. Where such costs equal or exceed the normal markup, the markup may be increased, and the amount of the increase depends on the contractor's judgment of the probabilities involved, his need for the particular job, and the relationship between the possible increased cost and his overall financial condition.

The bare cost of a completed project will never exactly equal its estimated bare cost, and there are no guarantees the contractor will make a profit on any job awarded to him. Because of the inherent uncertainties of the estimating process, there are practical lower limits to the amount of the markup used; these limits should not be violated often, if the contractor plans to stay in business. A theoretical analysis of past bidding history may indicate more money could be made by reducing the markup and thereby doing more work. The contractor, however, runs considerable risk when the markup used is below the practical lower limit for the type of work involved.

In the final analysis, the size of the markup will depend on the contractor's feeling for the job, and this feeling will be influenced by a number of considerations. Included among these will be his need for the particular job at the time, the amount of other work available, the number of other bidders, his past bidding experience on similar work against particular competitors, his past cost history on similar work, the type, size, and location of the job, the risks involved, and the character of the owner and his representative based on past experience, or inferred from the plans and specifications. In all cases the markup used should be considered both as a lump sum and on a dollar-per-month-of-job-duration basis.

In our estimates we will allow 10 percent for general overhead and profit on engineering work.

On building work the general contractor does about 25 to 30 percent of the work with his own forces, the uncertainties are fewer, and the markup is smaller. As rules of thumb, the following three parameters may be considered in determining a reasonable markup for building jobs.

1. 25 percent of the total bare cost of the general contractor's own work. This cost is taken directly from the estimate summary sheet and includes the contractor's costs for labor, payroll expenses, materials, and rentals. The costs for direct job overhead are included.
2. 40 percent of the total labor costs for the general contractor's own forces. This cost includes direct job overhead labor and all payroll expenses, and is the sum of the labor costs shown on the estimate sheets for the contractor's own items of work.
3. 5 to 7 percent of the estimated bare cost of the job. The bare job cost equals the total job cost, less the contractor's markup for general overhead and profit, and is shown on the estimate summary sheet. The upper limit is considered when the general contractor will do a larger percentage of the work with his own forces, as is usual when concrete buildings are involved. The lower limit is considered when more of the work is done by subcontractors, when the contractor has good cost records for the major categories of work required, and when no particular hazards are foreseen.

4

Structure Concrete—Engineering Work

The quantities of all materials, labor, and equipment that apply to the structure concrete item as a whole, together with the quantities of certain other minor activities, are collected and listed at the beginning of the estimate sheet for the structure concrete item in a section designated "General Costs." The costs of some of these items depend upon the time required to complete the structure concrete work, but none of them are normally on the critical path for the structure concrete item.

Separating the general costs from those labor and equipment costs directly related to performance of the work simplifies the estimate and the monitoring of costs on work in progress, decreases the possibility of error, and facilitates determining the time required for completion.

The separate operations required to complete the structure concrete item together with the quantities of their related primary activities are then listed below the general costs section in a logical sequence like that illustrated in the estimates that follow. The activities so listed are priced and the time required to complete the item is determined.

With the time required to complete the structure concrete item established, the general costs section is priced and the total estimated cost of the structure concrete item is obtained.

On some jobs there may be more than one job item for structure concrete. In such cases, a single combined take-off is made for all the structure concrete work and a single cost is determined for the total quantity of concrete involved. This procedure is necessary because many of the general costs relate to the total quantity of structure concrete on the project and cannot be realistically assigned to a particular bid item. When the total cost of the combined structure concrete items has been determined, the unit prices bid for the individual items are assigned somewhat arbitrarily, as illustrated in the estimates that follow.

4.1 General Costs Checklist

Those items from the following list that are applicable to the given job are listed and priced in the general costs section of the structure concrete estimate. Such listing and pricing requires very little time when a good take-off of the structure concrete item has been made.

1. Concrete.
2. Form lumber.
3. False work lumber.
4. Steel shoring.
5. Steel beams.
6. Column forms.
7. Concrete hoist.
8. Scaffolds and staging.
9. Form ties.
10. Minor expenses.
11. Make forms.
12. Additional finishing.
13. Additional curing.
14. Sandblasting.
15. Set anchor bolts.
16. Dry pack anchor bolts.
17. Compressor.
18. Vibrators.
19. Table saw.
20. Finishing machine.
21. Develop power.
22. Develop water.
23. Hoppers and buggies.
24. Neoprene-bearing pads and expansion joint material.
25. Curing paper.
26. Curing blankets.

27. Miscellaneous iron.
28. Deck and screed hardware.
29. Box girder dobies.
30. Clean girders.
31. Sweep bridge decks.
32. Clean up and out.
33. On and off charges for steel falsework.
34. Other.

4.1.1   Concrete

This item covers the purchase price, FOB job, of the portland cement concrete required. In the estimates that follow it is assumed that the concrete is available in transit-mix trucks.

An allowance of 2 percent for waste is adequate for all formed concrete and for most jobs taken as a whole. Where grade slabs constitute a large portion of the work on allowance of 5 percent for waste should be used for the grade slabs. Cast-in-drilled-hole piles require an allowance of 7 to 10 percent for waste.

4.1.2   Form Lumber

This item covers the purchase price, FOB job, of the form lumber and plyform required. The amounts required depend on the number of re-uses that are possible. Estimation of form lumber requirements is based on the concrete take-off, and is illustrated in the estimates that follow.

Wall form panels are assumed to be made with 2 × 4 studs and double 2 × 4 wales with form ties at 2 feet on center both ways. Plyform is assumed to be ⅝ inches thick. Studs are assumed to be 12 inches center to center except that, for walls or girders under 5 feet high, they may be on 16-inch centers.

In our estimates we will allow 1.75 to 2 MFBM of 2 × 4 lumber and 1.17 MSF of plyform for each MSF of wall form panels required.

When contractor-owned used material is available, it will be charged to the job at 1/3 the cost of new material, and this charge will include the cost of moving the material to the job.

4.1.3   Falsework Lumber

This item covers the purchase price, FOB job, of the falsework lumber required. When contractor-owned used falsework lumber is available, it will also be charged to the job at ⅓ the cost of new material, and this charge will include moving it to the job.

The amount of material required is based on a preliminary design for the falsework in all cases. Where large structural slabs with various arrangements of beams and girders occur, the falsework required for a typical bay is determined, and the total falsework required is estimated from the ratio of the total area to be formed to the area of the typical bay.

4.1.3.1 FALSEWORK DESIGN. In our estimates we will use the following loads and working stresses when designing falsework.

The minimum total design load will not be less than 100 pounds per square foot regardless of slab thickness.

Subject to the above minimum, the design load will be the greater of the loads determined by the following two methods:

1. 160 pounds per cubic foot for concrete plus the actual weight of any equipment supported plus 20 pounds per square foot plus 75 pounds per linear foot applied at the outside edge of deck overhangs.
2. 150 pounds per cubic foot for concrete plus the actual weight of any equipment supported plus 50 pounds per square foot plus 75 pounds per linear foot applied at the outside edge of deck overhangs.

Horizontal loading will be the greater of the wind load or 2 percent of the total dead load.

Deflection will be limited to 1/240 of the span.

The maximum stresses will be as follows:

Timber:

Compression perpendicular to the grain	450 psi
Compression parallel to the grain	$\dfrac{480,000}{(l/d)^2}$ psi
but not to exceed 1600 psi	
l = unsupported length of column in inches	
d = least dimension of column in inches	
Flexural stress	1500 psi
Horizontal shear	140 psi

In calculating horizontal shear stresses, all loads within a distance from either support equal to the depth of the beam will be neglected.

Nailed joints are common in constructing formwork. Whenever possible scaffold nails are used since they have a double head and are easily removed during the stripping operation. Nail sizes are designated by pennyweight which is abbreviated as d. For a given pennyweight, box nails are smaller in diameter than common wire nails, furnish less lateral

Table 4.1.   Properties of Wire Nails

Bright Common Nails			
Size (d)	Length (inches)	Wire Gage Number	Approximate Number per Pound
4	1½	12½	314
6	2	11½	181
8	2½	10¼	106
16	3½	8	49
Bright Scaffold Nails[a]			
6	1¾	11½	150
8	2¼	10¼	88
16	3	8	44

[a]The lengths shown for scaffold nails are the net length below the lower head.

resistance, and are not normally used in construction, since they bend easily when driven. Nails may be bright, galvanized, or cement coated.

The properties of wire nails are given in table 4.1. In our designs we will use 8d nails for fastening surfaced 1-inch material and 16d nails for surfaced 2-inch material. Assuming scaffold nails driven in the side grain of coast region douglas fir, we will use the following lateral resistances.

8d scaffold nail   70 pounds each

16d scaffold nail   100 pounds each

Steel:

All structural steel falsework will be designed in accordance with the Manual of Steel Construction published by the AISC.

4.1.4   Steel Shoring

This item covers the rental cost of heavy duty steel shoring similar to that shown in Figure 4.1. Such shoring is frequently used for bridge falsework. Under normal conditions, the cost to set and strip steel shoring is less than the cost to set and strip wood falsework bents. In addition, the danger of a falsework failure due to undetected defects in rough wood posts is eliminated.

The load-carrying capacity and the weight of the shoring towers must be obtained from the manufacturer in all cases. In our designs we will assume an allowable load of 10 kips per leg, and our setting and stripping

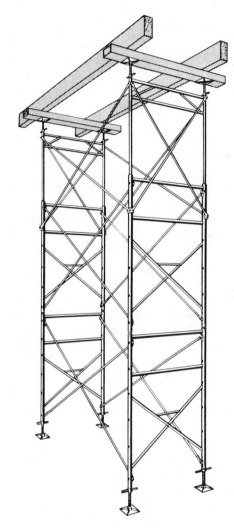

Figure 4.1 Shore "X" Tower. The trade-mark Shore "X" is registered by Waco Scaffold & Shoring Co.

costs will be based on a unit price per CWT for shoring of a given height. We will assume such shoring is contractor-owned, and will charge the jobs a monthly rental rate of $2 per CWT of shoring for those months the shoring is in use.

4.1.5   Steel Beams

This item covers the rental charged for steel wide flange beams used for falsework. In our estimates we will assume the beams are owned by the contractor, and we will charge a rental of $1.50 per CWT per use.

4.1.6   Column Forms

This item covers the purchase price, FOB job, of sonotube and the rental charged for steel column forms.

4.1.7   Concrete Hoist

This item covers the monthly rental charges for concrete hoists or conveyors, and includes the costs of moving the equipment on and off the job, and of operating and maintaining the equipment. Concrete hoists are used primarily on building work, and are discussed in Chapter 5. The hourly rental costs for truck-mounted concrete pumps or truck cranes used to place concrete are carried in the operating section of the estimate sheet, as illustrated in the following estimates.

4.1.8   Scaffolds and Staging

This item covers the monthly rental charges for portable scaffolds, scaffold jacks, and light weight metal staging, as well as the material costs for lumber, when wooden staging is used. The cost to erect and dismantle the staging is carried in the section of the estimate requiring the staging.

4.1.9   Form Ties

This item covers the purchase price, FOB job, and/or the rental charged for wall form ties. Our estimates are based on using contractor-owned taper ties at 2 feet on center both ways. We will charge the jobs a monthly rental rate for those months that the ties are being used. The monthly rental rate will be equal to or less than 1/12 the cost of a new tie, and the maximum rental charged on any job will never exceed the purchase price of new ties. Typical form ties are shown in Figure 4.37.

4.1.10   Minor Expenses

This item includes the costs of a number of items common to all concrete work. Included are the costs of nails, hand tools and incidental supplies, electric hand tools, saw filing, curing compound, form oil, and millwork. Individually, none of these items is of major importance, but as a whole they are significant, and an allowance must be made on the estimate for their cost. It is doubtful that any two contractors handle these items in exactly the same manner.

These items represent the costs for materials and services and each of them may vary considerably from job to job. They do not relate directly to wage rate indexes, nor to any of the general economic indexes available. In practice, where recent job costs records are available, the group as a whole may be bid with sufficient accuracy at a unit cost per MCY of concrete involved. The unit costs developed will vary with the type of work being done.

Since the student will not have access to a file of recent job costs, we will discuss each item individually, and derive a unit cost for the group based upon assumed current prices and the conditions involved on the jobs we bid in the following sections.

Nails.   This item covers the cost, FOB job, of the nails required to make, set, and strip the forms. The quantity of nails required per MCY of concrete varies with the number of re-uses of the form panels, the complexity of the job, the relative amounts of wall, slab, and foundation concrete, and with the thickness or mass of the structural sections. In our estimates we will allow 10.5 CWT of nails per MCY of concrete on engineering work, and 24 CWT of nails per MCY of concrete on concrete building work.

Hand Tools and Incidental Supplies.   This item covers the costs of hand tools, such as shovels and pry-bars, together with minor supplies such as curb stakes, rubber boots, corks, and number 9 wire. In practice, many of these items are obtained from the contractor's yard, and the job is only charged for the new ones purchased.

In our estimates we will use a cost of $500 per MCY for this item, which relates roughly to a wholesale price index of 200 (year 1967=100).

Electric Hand Tools.   This item covers the rental charged to the job for electric saws, drills, grinders, impact drills, and the like, that are required, together with the electric cords required for their use. In our estimates we will charge a monthly rental of 1/10 the cost of the tools involved, and will assume 2 saws and 1/2 drill will be required by each 6-carpenter crew.

Saw Filing.   This item covers the cost of filing the carpenter's hand saws as well as the electric saw blades. In our estimates, we will assume each 6-carpenter crew will require 3 handsaws and 2 electric saw blades to be sharpened for each week the crew works, and that the cost to sharpen either one will be 1/5 of the hourly wage rate for a carpenter. It is assumed that the filing will be done by a saw filing shop.

Curing Compound.   This item covers the cost, FOB job, of the curing compound and the sprays required to apply it. The labor cost to apply the curing compound is included in the cost of pouring the concrete. In our estimates, we will plan to use garden type hand spray cans, and will allow 1 gallon of curing compound for each 150 square feet of surface to be sprayed.

Form Oil.   This item covers the cost, FOB job, of the form oil and the sprays required to apply it. The labor cost to apply form oil and the labor cost to clean forms for re-use are included in the costs of making, setting, and stripping forms. In our estimates we will plan to use hand spray cans as before, and will allow 1 gallon of form oil for each 150 square feet of formed surfaces.

Millwork.   This item covers the cost, FOB job, of 3/4 inch chamfer strip and of special millwork items such as wedges or form marking strips when required. In our estimates we may take off the chamfer strip with the concrete, and will charge it at $0.13 per linear foot for the material. The labor cost to apply chamfer strip is included in the cost to set and strip forms. If marking strips are required on the form panels, a labor cost of $0.15 per linear foot of marking strip should be used, based on $K_{6+2} = 1000$, and this labor cost should be included with the cost to make the form panels.

4.1.11   Make Forms

This item covers the cost to make the required wall-form panels. The amount or area of forms to be made is determined on the concrete take-off, as shown in Chapters 4 and 5. Form marking strips, when required, represent an additional expense, as noted in Section 4.1.10.

4.1.12   Additional Finishing

This item covers the costs for the labor and materials required to fill form tie holes, remove fins, patch rock pockets, and grind or sack surfaces. The amount of additional finishing required will depend on the quality of formwork and on the job specifications.

   The total costs for additional finishing are fairly uniform for a given type of structure, and are generally proportional to the amount of concrete involved. In our estimates we assume first class formwork is used, and we estimate the cost of additional finishing based on a unit cost per MCY of concrete in a particular type of structure in many cases. When necessary,

Table 4.2. Unit Costs for Additional Finishing Work

Activity	Unit Cost for Labor	Index	
Fill form tie holes	$600–$650 per M holes each face[a]	K_{L+F}	= 29.58
or	$300 per M holes each face[b]	K_L	= 13.48
Grind for uniform			
color	$270 per MSF	K_F	= 16.10
Sack nearby walls	$500–$575 per MSF	K_F	= 16.10
Sack beam sides,			
and beam and			
slab soffits	$260 per MSF	K_F	= 16.10
Touch up and patch	$30 per MSF formed[c]	K_F	= 16.10

[a]Exposed, nearby wall surfaces. White cement is used, and patches are not visible on completion.
[b]Unexposed surfaces. Patches visible upon close inspection.
[c]Exclude soffit areas.
Note: For estimating purposes, assume the costs for materials equal 4% of the cost for labor.

the quantities involved will be taken off and the unit costs presented in Table 4.2 will be used to determine the cost of the additional finishing.

When the soffit formwork for structural slabs and girders has been properly made, very little touch-up is required for the soffits. Therefore the areas of structural slab and girder soffits are not included in the square footage used to estimate the costs to touch up and patch.

The cost to fill form-tie holes constitutes a substantial portion of the additional finishing cost. It is important that no costs be charged where form-tie holes do not need to be filled, as may be the case in connection with some foundation walls and the interior girders of box girder bridges.

4.1.13 Additional Curing

This item covers the costs to water cure concrete on weekends and holidays. The normal cost for curing on working days is included in the cost to pour the concrete. In the rare case where the specifications require one or more men to be assigned to curing on a full-time basis, the entire cost for such labor should be included under this item.

4.1.14 Sandblasting

This item covers the labor and material costs for sandblasting construction joints. In our estimates, we will use a unit cost per MCY of concrete in some cases and a unit cost per square foot in others.

Where the areas to be sandblasted at any one time range from 50 to 800 square feet, we will use a unit cost for sandblasting labor of $0.30 per SF, indexed to $K_L = 13.48$. The sand used must be dry. We will assume it is purchased in 100 pound sacks, and that 150 pounds of sand will be used per hour. The rental cost for the sandblast pot, hose, and nozzle will be included in the rental charged for the compressor.

The cost to water-cut laitance from freshly poured construction joints is included in the unit costs to pour concrete.

4.1.15   Set Anchor Bolts

This item covers the labor cost to set anchor bolts for heavy building columns or for bridge girder connections. Such bolts may be 1-1/4 inches in diameter by 24 inches long, will usually have 2 or 4 bolts in each set, and will often be furnished with plate washers and pipe sleeves.

In our estimates we will use a unit cost for labor of $24 per 2-bolt set, and $44 per 4-bolt set, based on $K_{6+2} = 1000$. These unit costs will include the costs to make, set, and strip the bolt templates, and to set and adjust the bolts.

The material cost of the anchor bolts is included in the cost of the job items for structural steel or miscellaneous iron in all cases.

4.1.16   Dry Pack Anchor Bolts

This item covers the cost to dry pack the anchor bolt pipe sleeves and the column base plates or masonry plates. We will use a unit cost of $50 per cubic foot to place dry packing, based on $K_{L+F} = 29.58$. In addition, an allowance for materials will be made, based on using 1 part of cement to 2-1/2 parts of concrete sand for the dry pack mix.

4.1.17   Compressor

This item covers the rental charges for a 105 to 120 CFM portable air compressor, and includes the charges for small air tools, air hose, and the sandblast pot.

4.1.18   Vibrators

This item covers the rental charges for the concrete vibrators. In our estimates we will plan to have a backup vibrator available for each pour. The labor cost for vibrating concrete is included in the costs for pouring concrete.

4.1.19   Table Saw

This item covers the rental charge for a table saw when required. The cost of the saw man is included in the cost of those activities requiring use of the table saw.

4.1.20   Finishing Machine

This item covers the rental charges for the rotary floating and troweling machine used on building slabs, and the deck finishing machine used on bridge decks. Deck finishing machines are normally contractor-owned. In our estimates we will use a rental rate of $0.05 per square foot of bridge deck to cover the cost of the deck finishing machine, including fuel and maintenance. The cost of the operator is included in the labor cost to pour the bridge deck.

4.1.21   Develop Power

This item covers the cost to develop power for use on the concrete work. On building work, it normally includes the cost of a job service pole, the connect and disconnect charges by the power company, and the cost of additional electric cords if necessary. Portable generators are often used on engineering work, since the power is required at a number of locations, and often for only short periods of time. In such cases a monthly rental charge is made for the generators required, and this rental charge includes the cost of fuel and of additional electric cords when required.

4.1.22   Develop Water

This item covers the costs to develop the water required for the concrete item, and includes the costs of the hoses and nozzles required to wet down the forms, cure the concrete, and water-cut laitance from fresh construction joints when required.

4.1.23   Hoppers and Buggies

This item covers the rental charges for concrete hoppers, elephant trunks, and concrete buggies when required.

4.1.24   Neoprene Bearing Pads and Expansion Joint Material

This item covers the cost, FOB job, to furnish neoprene bearing pads used on certain bridges, and expansion joint material.

4.1.25 Curing Paper

This item covers the cost, FOB job, for curing paper used on interior building slabs. On such work, paper is required to protect the slab even though curing compound is used. When taking off the paper required, an allowance must be made for the overlaps at the sides. The labor cost to apply such paper on building slabs is included in the costs to pour and cure the slabs.

4.1.26 Curing Blankets

This item covers the cost, FOB job, to furnish the curing blankets, burlap, or polyethylene required to water-cure structural concrete. Used rugs are often used to water-cure bridge decks, and are sold by the ton. The labor cost to place and move this material is included in the cost to pour and cure the concrete.

4.1.27 Miscellaneous Iron

This item covers the cost, FOB job, of the miscellaneous iron required for the contractor's own use in connection with the formwork. Such material is not covered in the structural steel or miscellaneous iron items for the job. For example, the tie rods used to secure out-riggers to the exterior girders in connection with the formwork for deck overhangs is included here. The labor cost for installing the miscellaneous iron is included with the cost of the formwork involved.

4.1.28 Deck and Screed Hardware

This item covers the cost, FOB job, of screed pipes and supports on building work and of the deck form hangers used on precast concrete or structural steel girder bridges. These costs could be included with the miscellaneous iron, but are listed separately to ensure that they are not overlooked.

4.1.29 Box Girder Dobies

This item covers the cost of the labor and material required to make the "dobies" used to support the interior girder and diaphragm forms on box girder bridges. These are concrete blocks 3-1/2 inches square, with a height equal to the depth of the bottom slab of the box girder bridge. In our estimates we will use a labor cost to make the dobies of $700 per MCY of concrete in the bottom slab and girder pour, based on $K_{6+2} = 1000$. In addition we will allow 1.5 CY of concrete per MCY of concrete in the bottom slab and girder pour, for the material required.

4.1.30 Clean Girders

When bridge decks are poured on steel girder bridges, the primed girder webs are inevitably streaked with water and cement. This streaking must be removed prior to applying the final coats of paint. In our estimates, we will allow a labor cost of $120 per MSF of girder web to be cleaned, based on $K_L = 13.48$.

4.1.31 Sweep Bridge Deck

Prior to obtaining final acceptance, the bridge decks must be swept clean. In our estimates we will charge a labor cost of $13 per MSF of deck to be swept, based on $K_L = 13.48$.

4.1.32 Clean Up and Out

This item covers the cost of moving all form materials off the job on completion. In our estimates we will charge a labor cost of $43 per MFBM or MSF of form lumber or plywood ordered for the job, based on $K_L = 13.48$. In addition we will charge a rental cost for trucks and for hoisting equipment if needed.

4.1.33 On and Off Charges for Steel Falsework

The costs of moving steel falsework on and off the job must be included when this material is rented from outside sources. In our estimates we will assume that the steel falsework is contractor-owned and that the cost of moving it onto the job is included in the rental charged. We will charge the cost to move the steel falsework off the job on completion. Based on $K_L = 13.48$, we will charge $34 per ton for labor and we will charge a rental cost for trucks and for hoisting equipment, if needed.

4.1.34 Other

This item will cover the costs of any requirements relating to the concrete item which are not covered in the costs listed above. For example, the costs for mix designs or testing, if required, would be included here.

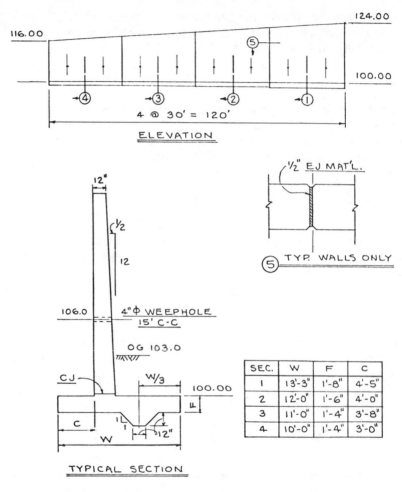

ELEVATION

TYPICAL SECTION

SEC.	W	F	C
1	13'-3"	1'-8"	4'-5"
2	12'-0"	1'-6"	4'-0"
3	11'-0"	1'-4"	3'-8"
4	10'-0"	1'-4"	3'-0"

NOTE: 2 WALLS REQ'D 1 AS SHOWN, 1 OPP. HAND

Figure 4.2 Retaining walls.

4.2 Retaining Walls

We will now prepare a complete bid for the retaining walls shown in Figure 4.2. Reinforcing steel is a subcontract item and no take-off for it is necessary. For clarity, reinforcing steel is not shown on this, nor on succeeding plans.

We will assume the job plans call for two retaining walls—one as shown in Figure 4.2, and one opposite hand. The two walls are to be constructed opposite each other, 50 feet to the right and left of the centerline of a future roadway. The engineer's estimate for the retaining walls is as follows:

ENGINEER'S ESTIMATE

Item No.	Item	Unit of Measure	Estimated Quantity
1	Clearing and Grubbing	LS	Lump Sum
2	Structure Excavation	CY	550
3	Structure Concrete	CY	425
4	Reinforcing Steel	LB	44000

The following job conditions are assumed:

1. Job is located 15 miles from the contractor's yard, and from a rental crane source.
2. Clearing and grubbing consists of 1 tree, 10 inches in diameter by 30 feet in height.
3. The earth is sandy loam.
4. There are no overhead obstructions.

Figure 4A Retaining walls, Truxtun Avenue Overpass. Contractor: Tumblin Company.

5. A water supply is available at a nearby residence.
6. No traffic control is required.

4.2.1 Structure Concrete

The structure concrete is the major job item; it is taken off and its cost is estimated first.

4.2.1.1 THE CONCRETE TAKE-OFF.

The concrete take-off is shown in Figures 4.3 and 4.4. The foundations and the walls are taken off separately, since they represent two different operations. The primary activities required appear in abbreviated form as the column headings, and their quantities are taken off and entered in the appropriate columns.

The S_1 totals summarize the foundation work, and show the CY of concrete, the contact area of foundation forms 1.5 feet in height, the SF to be fine graded, and the CY of hand excavation required for the slab key. The area to be fine graded is the contact area of all concrete with the subgrade. The miscellaneous column is used to collect additional items that may be unrelated to each other but which must be included in the estimate. In this instance, we note the gallons of curing compound required to cure the top face of the foundation slab.

Similarly, the S_2 totals summarize the wall work and show the CY of concrete and the contact areas of wall forms to be set and stripped. The wall areas are segregated by height, and include the areas of the end bulkheads. The area to be sandblasted, the required amount of expansion joint material and chamfer strip, and the number of weep holes are noted in the miscellaneous column.

The total quantity of concrete in the retaining walls is found to be 423.0 CY, which is in reasonable agreement with the 425 CY shown on the engineer's estimate.

The plans indicate a starter wall at the construction joint. It is not necessary to consider this in the estimate, since the costs to set and strip starter walls and solid keys are included in the unit cost to set and strip foundations.

The take-off for the general cost items is based on the quantities listed in sections S_1 and S_2. We decide to furnish foundation forms for one foundation complete, and to start setting wall forms the day after the first foundation pour. On this basis, wall forms must be furnished for the three larger wall pours, for one set of walls, to keep the crew from running out of work.

The foundation forms will be made of 2-inch material from the contractor's yard. The area from S_1 is converted to MFBM and is listed as

RETAINING WALLS	MISC.	H EX	FG	H 15	CY
FOUNDATIONS					
$13.25 \times 30 \times 1.67/27$			398		24.6
74.5×1.67				125	
$2.00 \times 1 \times 30/27$		2.2			2.2
$12.00 \times 30 \times 1.50/27$			360	92	20.0
61×1.50					
$2.00 \times 1 \times 30/27$		2.2			2.2
$11.00 \times 30 \times 1.33/27$			330		16.3
61×1.33		2.2		82	2.2
$10.00 \times 30 \times 1.33/27$			300		14.8
70×1.33		2.2		93	2.3
1 ADD'L. FND.		8.8	1388	392	84.6
CURE COMP. $2776/150 = 19$ GAL.	1				
S_1	1	17.6	2776	784	169.2

WALLS	MISC.	H 17	H 19	H 21	H 23	CY
$A_1 = (1+24\times0.021)24 = 36.1$					36	
$A_2 = (1+22\times0.021)22 = 32.2$					33	
$68.3/2 \times 30/27$						37.9
$23 \times 2 \times 30$					1380	
$A_3 = (1+20\times0.021)20 = 28.4$				29		
$60.6/2 \times 30/27$						33.7
$21 \times 2 \times 30$				1260		
$A_4 = (1+18\times0.021)18 = 24.8$			25			
$53.2/2 \times 30/27$						29.6
$19 \times 2 \times 30$			1140			
$A_5 = (1+16\times0.021)16 = 21.4$		22				
$46.2/2 \times 30/27$						25.7
$17 \times 2 \times 30$		1020				
4 WALLS ADD'L.		1042	1165	1289	1449	126.9
SAND BLAST $1.83 \times 240 = 440$ SF	1					
$\frac{1}{2}"$ EJ MAT. $20 \times 2 \times 6 = 240$ SF	1					
CHAMFER $(20+1+20)16$						
$+4 \times 120 = 1136$ LF	1					
WEEP HOLES 16 EA. $4"\phi \times 2'$	1					
S_2	4	2084	2330	2578	2898	253.8

$\Sigma_1^2 S = 423.0$ CY

Figure 4.3 Concrete take-off for retaining walls.

RETAINING WALLS – CONT'D.	MISC.	Σ SF	SF MAKE	MSF PLY	MFBM 2×4	MFBM 2× HAVE
GENERAL COSTS						
S₁ FURNISH 400 SF		784				0.8
STAKES & BRACES						0.3
S₂ MAKE 1165+1289+1449=3903		9890	3903			
3.90 × 1.17				4.6		
3.90 × 2					7.8	
GRIND 18×120×2 = 4320 SF	1					
HOLES 9600/4 = 2400 TOT.	1					
EXPOSED $\frac{2400}{2} - \frac{240}{2} = 1080$						
TAPER TIES 3903/8 = 488 EA.	1					
FORM OIL 10674/150 = 71 GAL.	1					
S₀	4	10674	3903	4.6	7.8	1.1
				HAVE 2.3		

Figure 4.4 Concrete take-off for retaining walls (cont.).

shown, together with an allowance for 2 × 4 stakes and braces. The unit costs for setting and stripping foundation forms include the cost of building the forms in place with 2-inch material; no additional allowance for making these forms is necessary.

The area of the wall form panels to be made is taken from section S_2, and is listed in the "SF Make" column. This area is converted into MFBM of 2 × 4 and MSF of plyform as shown.

Since we plan to bid the additional finishing cost on a unit cost basis, the total contact area of forms is taken from S_1 and S_2 and is listed in the "ΣSF" column. The area to be ground is based on an assumed requirement of the specifications, and includes all the exposed front wall areas above elevation 102.

The wall form ties are assumed to be set on 2 foot centers, and the number of form tie holes to be filled is determined by dividing the total area of formed wall faces, from section S_2, by 4 SF per hole. This number is divided into exposed and unexposed holes, since the costs differ. For this calculation it is assumed that on completion the bottom row of holes on the front wall faces will be covered.

The number of taper ties required is found by dividing the area of wall form panels to be made by 8 SF per tie. Taper ties are completely reuseable. If snap ties or she-bolts were to be used, an allowance would be made for the cost of the embedded and lost parts, and this allowance would be based on the total area of all of the walls.

A note is made at the bottom of the "MSF Ply" column that we have 2.3 MSF of used plyform in our yard.

4.2.1.2   THE STRUCTURE CONCRETE ESTIMATE.   The estimate sheet for the concrete item is shown in Figure 4.5. In all cases the applicable wage and rental rates are listed at the top of the sheet. Those items from the general cost check list that are applicable are listed under the S_0 heading. The S_1 and S_2 items are listed directly from the concrete take-off.

The unit costs relating to sections S_1 and S_2, together with the indexes on which they are based, are given in Table 4.3 and Figure 4.6. The unit costs used for the additional finishing items are shown in Table 4.2. The estimated cost of doing the work involved in sections S_1 and S_2 is determined, and an estimate of the time of completion for the item is made.

The estimated time to complete each section is determined in the days column of the estimate sheet. The total labor cost, less the pour labor cost for section S_1 and for section S_2, is divided by the daily cost of the proposed crew to obtain the required number of working days to set and strip the formwork. A four-carpenter crew is assumed for the foundations, and a six-carpenter crew is assumed for the walls.

ITEM 3. 425 CY STRUCTURE CONC.
(423.0)

C 16.34 K₆₊₂ 1000 / L 13.48 / F 16.10 Kcr 75		L	DAYS	M,R

Top-left header block:
$$C\ 16.34 \qquad K_{6+2}\ 1000$$
$$L\ 13.48$$
$$F\ 16.10 \qquad K_{cr}\ 75$$

		L	DAYS	M,R
S₀ GENERAL COSTS				
CONC. 423.0 ×1.02 = 432 CY	30			12960
LUMBER BUY 7.8 MFBM	300			2340
✓ HAVE 1.1 ✓	100			110
PLY 2.3 MFS @ 500 + 2.3 MSF @ 170				1540
TIES 4.9 C 1 MO.	40/C			195
MINOR EXP. 0.423 MCY	1950			825
MAKE FORMS 3900 SF	0.22	860		—
ADD'L. FINISH. 0.423 MCY	5980	2530		100
✓ CURE 5 WKS.	50	250		—
SAND BLAST 440 SF	0.30	130		20
COMPRESSOR 1 MO.	400			400
VIBRATORS (2) 1 MO.	200			200
GENERATOR 1 MO.	250			250
DEVELOP WATER 50+100				150
EJ MAT'L. 240 SF (½") 0.08	0.50	20		120
BURLAP 4 MSF		—		100
CLEAN UP 13.5 MFBM 45	12	580		160
S₁ FOUNDATIONS			2130	
18 CY H. EX.	14	250	1000×⅔	
2780 SF FG	0.14	390	=4,+1	
SS 784 SF H1E	1.90	1490	5D	
POUR 173 CY	6.92	1195		
1692				
S₂ WALLS			12520	
WEEP HOLES 16 EA @ 8 32 LF @ 2		130	1000	65
SS 2084 SF H17	1.20	2500	=13,+8	
SS 2330 ✓ H19	1.23	2870	21D	
SS 2578 ✓ H21	1.26	3250		
SS 2898 ✓ H23	1.30	3770		
POUR 259 CY	6.10 5.90	1580		1530
2530 / 4230		21795		21065
INCL. PAY ROLL EXP (1.195)		26045		
$\Sigma = \$47,110$ ($111/CY)			3+21 = 24D	

MINOR EXP.

NAILS 10.5×0.423 = 4.4 CWT	40	
HAND TOOLS 0.423 MCY	500	
ELEC. TOOLS 1 MO.	70	$825
FILING 5 WKS	17	
CURE COMP. 19 GAL @1, +40		
FORM OIL 71 GAL @ 0.50, +40		
MILLWORK 1136 LF	0.13	

ADD'L FIN.

TOUCH UP	10.67 MSF	30	
HOLES	1.08 M	600	$2530
	1.32 M	300	
GRIND	4.32 MSF	270	

Figure 4.5 Estimate for retaining wall concrete.

Table 4.3. Unit Costs for Retaining Walls

Activity	Labor		Rental		Unit of Measure
	Unit Cost	Index	Unit Cost	Index	
Foundations					
Fine grade	$ 0.14	$K_L = 13.48$	—		SF
Hand excavation	$14.00	$K_L = 13.48$	—		CY
SS H 1 – H 3	$ 1.90	$K_{6+2} = 1000$	—		SF
Pour foundations	$ 3.72	$K_L = 13.48$	$3.70	$K_{Cr} = 75$	CY
or	$ 6.92	$K_L = 13.48$	—		CY
Walls					
Make form panels	$ 0.22	$K_{6+2} = 1000$	—		SF
Pour walls	$ 6.10	$K_L = 13.48$	$5.90	$K_{Cr} = 75$	CY

Note: When pours are large or average in size, the unit costs for the crane include 1 hour of travel time. On small pours, the travel time is additional, and travel time in excess of 1 hour should be added in all cases. Unit costs to set and strip retaining walls are given in Figure 4.6.

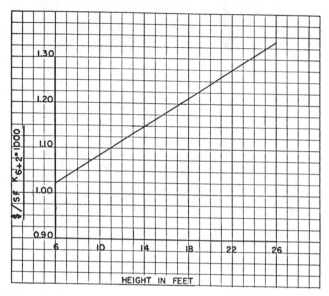

Figure 4.6 Unit labor costs to set and strip retaining wall forms when total area = 5000 SF or more. Increase unit costs 25% for areas of 1000 SF.

The cost of the pour labor is not included, since only the first foundation pour and the last wall pour are on the critical path for the item. On larger jobs, the concrete work is normally completed well before the end of the job, and all pour times are neglected; those that are on the critical path for the item will be small compared with the times required for the total formwork, and will be within the limits of accuracy for the time estimate.

In our case, the total time required is small, and the concrete work is essentially the entire job. Therefore, 1 working day is added to the foundation time to allow for the first foundation pour and 8 working days are added to the wall time; 1 day is allowed for the last wall pour, 2 days are allowed for curing with forms in place, and 5 days are allowed for completing the cure and for the additional finishing. It should be remembered that the time required to strip forms is included in the time required for the formwork, and that there are a minimum of 7 curing or calendar days for each 5 working days.

Since half the foundation work will be overlapped with the wall work, and since all of the wall work is on the critical path, the total time to complete the concrete work will be 3 + 21 = 24 working days.

With the time for completion established, the items previously listed in the general costs section are priced. In practice, all the material and outside rental costs are based on quotations from the suppliers or on recent invoices. We have assumed that one of the laborers will monitor

Figure 4B Retaining Walls and Steal Beam Bridge, Chester Avenue Underpass. Contractor: Tumblin Company.

the weekend cure for $50 per weekend, that $50 will buy the water from the home owner, and that an additional $100 will cover the costs of hoses and nozzles. Although the cost to place incidental amounts of expansion joint material is included in the costs of the formwork, we have allowed an additional $0.08 per square foot to place this material because an unusual amount of layout is required.

The cost columns for labor and for equipment rental are totaled, and an allowance of 19.5 percent is added to the total labor cost for payroll expenses. This percentage is based on workmen's compensation insurance rates for concrete work NOC, and the method of determining the allowance for payroll expenses is discussed in Chapter 3.

The total cost for the item is entered on the estimate summary sheet illustrated in Figure 4.8.

4.2.2 Structure Excavation

We will assume that the specifications permit the structure excavation to be disposed of at a nearby location having an average haul distance of 250 feet one way. We will also assume that the horizontal pay limits for structure excavation are specified to be vertical planes one foot outside the neat lines of the footings. The estimate sheet for the structure excavation item is shown in Figure 4.7.

ITEM 2. 550 CY STR. EX.
(570)

5 CY RT LOADER $60/HR.	L	DAYS	M,R
LOADER ON-OFF 2 HRS 60 782 CY (TRUE) 0.50 Σ = $ 510 LOAD & DUMP 40 SEC 0.67 MIN. TRAVEL 500 LF/370 FPM 1.35 2.02 MIN. 55/2.02 × 5 × 0.9 = 122 CY/HR 782/122 = 6.4 SAY 6.5 HRS. 6.5 × 60/782 = $ 0.50/CY	— —	$\frac{390}{60 \times 8}$ = 1 D	120 390 510
DIRECT JOB OVERHEAD BONDS 70 M @ 7.50 SUP'T. 6 WKS @ 550 RADIO 1.5 MO. @ 70 TOILET ✓ @ 60 ICE & CUPS VEHICLES (2) 3 V. MOS. @ 280 INCL. PAYROLL EXP. (1.195) Σ = $ 5555 TIME EST. MOVE ON-OFF 2 D STR. EX 1 CONCRETE 24 REINF 3 30 WORKING DAYS	3300 3300 3945		— — 525 — 105 90 50 840 1610

Figure 4.7 Estimates for structure excavation and direct job overhead.

The pay quantity of structure excavation shown on the engineer's estimate is checked and found to be 570 CY, including the 18 CY of hand excavation that was taken off and priced in section S_1 of the concrete estimate. The 570 CY figure is written in parenthesis under the engineer's estimated 550 CY at the top of the sheet, and we note that the item should overrun the engineer's estimate. The actual structure excavation based on 1:1 side slopes is found to be 782CY. For this calculation the bottoms of the trapezoidal cross sections used are approximately the width of the concrete footing plus 1 foot on each side. The estimate for structure excavation is based on using a wheel loader with a 5 CY bucket. The calculation of the loader time required is shown on the estimate sheet. The loading and dumping time of 40 seconds takes into account the shallow cut and the need to make a rough grade 0.1 foot above finish grade. The average travel speed for the loader is taken at 370 feet per minute, and a 55-minute hour is used to compensate for minor delays. In addition, it is assumed that the average bank yards carried per bucket equal 0.9 of its rated capacity of 5 CY. A detailed discussion of excavation with wheel loaders is included in Chapter 7.

Since the structure excavation will be completed in 1 day at the start of the job, we assume that the required grade checking will be done by the superintendent, and no additional charge is made for grade checking labor.

4.2.3 Clearing and Grubbing

Clearing and grubbing costs are discussed in Chapter 11. The item is included here to illustrate how the contractor's direct job overhead and his markup are distributed among the unit prices bid for the work.

For this job we assume that a local tree service will dispose of the tree for $80 after we have uprooted it with the loader used for the structure excavation, and we allow 1/2 hour additional loader time for this. Since only two figures are involved, the cost information is entered on the summary sheet in the double space below the listed item.

4.2.4 Direct Job Overhead

The direct job overhead estimate is also shown in Figure 4.7. The items listed are taken from the direct job overhead check list presented in Chapter 3. The bond cost is based on an approximate estimate of $70,000 for the bid price of the job, and the percentage used for payroll expenses applicable to the superintendents salary is determined by the methods set forth in Chapter 3. The time estimate for the job is shown at the bottom of the direct job overhead sheet, and is used to determine the cost of the superintendent.

4.2.5   The Estimate Summary Sheet

The estimate summary sheet is shown in Figure 4.8. The items from the engineer's estimate are listed, and the quantities checked against those shown on the bid proposal. The costs for the contractor's own work including his direct job overhead are taken from the estimate sheets and listed in the cost column as shown. The unit price quoted by the low reinforcing steel subcontractor is extended and entered in the cost column. The column is totaled, 10 percent is added for general overhead and profit, and the total job cost is obtained. The cost previously used to calculate the bond cost on the direct overhead sheet is reviewed and found to be correct.

It is now necessary to assign unit prices to the various items in such a manner that all the direct job overhead and all the allowance for overhead and profit will be paid on completion, and that the total of the price column will be in reasonable agreement with the total of the cost column.

Reinforcing steel is a subcontract item, and we are not sure that the quantity shown on the engineer's estimate is correct. We therefore list a unit price for reinforcing steel equal to the low bid received, rounded off to even cents. Thus, even though the item might underrun by a substantial amount, we will not lose any of our overhead and profit costs on the item.

We arbitrarily list $3000 as the lump sum price for clear and grub. This is less than 5 percent of the total job cost, and is assumed to be in line with past bids for similar work in the area. The price of structure excavation is arbitrarily set at $3 per CY, since we expect the item to overrun the engineer's estimate.

A trial determination of that unit price for concrete that will exactly balance the price bid with the total cost is made, and the balancing price is found to be $129.44 per CY. Since we prefer to bid our major item to an even dollar, we enter a unit price of $130 per CY for structure concrete, and decrease the unit price bid for structure excavation to $2.50 per CY.

If the quantity shown on the engineer's estimate for structure concrete had been overstated by an appreciable amount, we would have been forced to bid the item at cost, which in this case would be $111 per CY, and to have balanced the cost-bid prices by increasing the prices bid for clear and grub and for structure excavation.

It should be noted that regardless of the price we bid for subcontract items, we will pay the subcontractors at the prices they quote to us in all cases.

SUMMARY RETAINING WALLS			COST		UNIT PRICE	TOTAL PRICE
1	CLEAR AND GRUB 80+30 = 110	LS	110		3000	3000
2	STRUCTURE EX.	550 CY	510		2.50	1375
3	STRUCTURE CONC.	425 CY	47110		130	55250
4	REINFORCING STEEL ACME 22.15/CWT	44000 LB	9750		0.22	9680
	DIRECT JOB OVERHEAD		5560			
			63040			
		+ 10%	6300			
			69340			69305

Figure 4.8 Estimate summary sheet for retaining walls.

4.3 Box Culverts

A complete bid for the box culvert shown in Figure 4.9 will now be made. The engineer's estimate is assumed to be as follows:

ENGINEER'S ESTIMATE

Item No.	Item	Unit of Measure	Estimated Quantity
1	Remove Bridge	LS	Lump Sum
2F	Structure Excavation	CY	230
3	Structure Concrete	CY	150
4	Reinforcing Steel	LB	22100

The job site inspection is assumed to reveal conditions similar to those given for the preceding retaining walls. No clearing and grubbing is required and the excavated material may be left anywhere on the site.

4.3.1 Structure Concrete

The structure concrete is the major job item, and its cost is estimated first.

4.3.1.1 THE CONCRETE TAKE-OFF.   The concrete take-off is shown in Figures 4.10 and 4.11. The culvert will require three pours. The headwall foundations, bottom slab, and cut-off walls will be poured first; the headwalls will be poured second; and the box walls and top slab will be poured last.

The primary activities required to build the structure are shown as column headings on the take-off sheets. Thus the S_3 section indicates that 60.0 CY of concrete are involved, that 177 SF of headwall foundation forms 1.2 feet high, 70 SF of slab side forms 0.6 feet high, and 265 SF of cut-off wall forms 4 feet high must be set and stripped; 2023 SF must be fine graded; and 2 CY of hand excavation are required.

The area listed for the cut-off walls is the area of the one formed face plus an allowance for additional formwork estimated to be required at the box end of the headwall foundations, in order to contain the box slab concrete in that region where the theoretically vertical bank will slough back during the excavation.

The miscellaneous item in the S_3 section covers the additional concrete estimated to be required in the above region and at the back face of the cut-off walls.

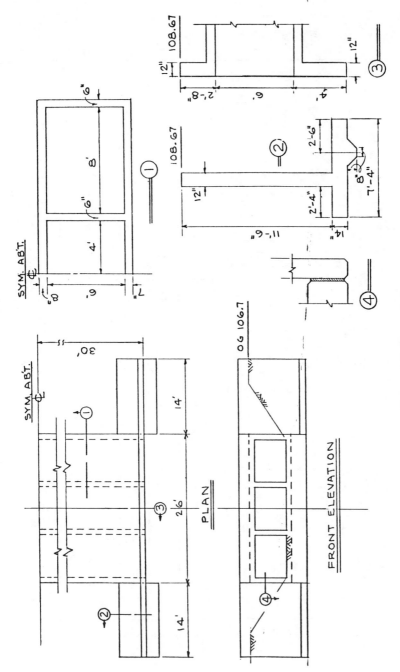

Figure 4.9 Box culvert.

BOX CULVERT — I.

	MISC.	H EX	FG	H4	HO⁶	H1²	CY
FOUNDATIONS· & BOT. SLAB.							
HEADWALL F'NDS.							
$(14 \times 7.33 \times 1.17/27)\,4$			411			177	17.8
$(28+7.33+2.33)\,1.17 \times 4$							
$1.33 \times 0.67 \times 14 \times 4/27$		1.9					1.9
BOT. SLAB							
$26 \times 60 \times 0.583/27$			1560				33.7
$2 \times 60 \times 0.583$					70		
CUT OFF WALLS							
$(3.42 \times 26 \times 1/27)\,2$			52				6.6
$4 \times 26 \times 2 + 2.3 \times 6.2 \pm \times 4$				265			
$\overline{0.9} \times 3.4\,(26 \times 2 + 5 \times 4)/27$							
$= 8$ CY ADD'L. CONC.	1						
S₃	1	2	2023	265	70	177	60.0

					MISC.	H12	CY
HEADWALLS.							
$11.5 \times 14 \times 1/27 \times 4$							23.9
$23 \times 14 \times 4 + (11.5 + 8.67) \times 1 \times 4$						1369	
48 SF EJ MAT'L.					1		
S₄					1	1369	23.9

					SOF	H2	H6	CY
BOX WALLS & TOP SLAB.								
BOX WALLS								
$0.5 \times 6 \times 60/27 \times 4$								26.7
$6.67 \times 58 \times 2 + 6 \times 60 \times 6$ $\}$							2958	
$\quad + 0.5 \times 6 \times 8$								
TOP SLAB								
$26 \times 60 \times 0.667/27$								38.5
$8 \times 60 \times 3$					1440			

Figure 4.10 Concrete take-off for box culvert.

BOX CULVERT   CONT'D.

		SOF	H2	H6	CY
PARAPET WALLS					
$1 \times 2 \times 26 \times {}^2/27$					3.9
$(2.67+2) \times 26 \times 2$			243		
S_5		1440	243	2958	69.1

$\Sigma_3^5 S = 153.0$ CY

		MISC.	SF MAKE	MSF PLY	MFBM 4x HAVE	MFBM 2x
GENERAL COSTS						
S_3	$512 \times 1.1 \times 2$					1.2
S_4	$(11.5+8.67)4$		81			
S_5	$2958+243$		3201			
	$3282 \times 1.17, 1.75$			3.8		5.7
	SOFFIT 1440×1.17			1.7		
	2×4 $150/8$ }					
	2×6 360 LF }					1.2
	4×6 $9/12$ $12/10$ }					
	4×4 $39/6$ }				0.7	
	$3201/8 = 400$ TAPER TIES	1				
MINOR EXPENSES	S_0	1	3282	5.5	0.7	8.1
				HAVE 4.0		

NAILS 10.5 x 0.153 = 2 CWT   40
HAND TOOLS 0.153 MCY   500
ELEC. TOOLS 1 MO.   70
FILING   3 WKS.   17   #530
CURE COMP.   21 GAL @1, +40
FORM OIL   44 GAL @ 0.50, +40
CHAMFER 870 LF   0.13
40 PAIR WEDGES @ 0.40

Figure 4.11 Concrete take-off for box culvert (continued).

Figure 4C Box Culvert, Central Branch Canal. Contractor: Tumblin Company.

The amount of form lumber required is determined in the general costs section. The total area of foundation forms, taken from S_3, is converted to MFBM of 2-inch material, allowing 10 percent for waste, braces, and stakes. The unit costs for setting and stripping foundation forms include the cost of building them in place with 2-inch material; no additional allowance for making these forms is needed.

The required areas of wall forms in sections S_4 and S_5 are reviewed, and we decide to furnish forms for section S_5 and use part of this material for the forms required for the headwalls in Section S_4. Accordingly, the only forms to be furnished for the headwalls are the end bulkheads. These are 1 foot wide and do not lend themselves to re-use. The conversion of the headwall forms to box wall forms is simple in this case. The headwall form panels will be made 12 feet high, and will be cut in half for use as box wall forms. If the conversion had required an appreciable amount of modification, an allowance based on the area to be modified, the extent of the modification, and a consideration of the cost to make the wall form panels would have been made and noted in the miscellaneous column.

The total area of wall form panels to be made is determined by adding all the wall areas in section S_5 to the area of the headwall bulkheads. This total area is converted to MFBM of 2 × 4 and MSF of plyform. The material required for the soffit form is based on a quick preliminary design, and we plan to use 2 × 4s 16 inches center to center, supported on 2 × 6 ledgers nailed to the box wall form studs at each end and supported by a 4 × 6 cap at midspan. The cap will be on 4 × 4 posts, spaced 5 feet center to center. We assume we have the 4 × 4s and 4 × 6s in our yard, and the amount required is listed in the "have" column.

4.3.1.2 THE STRUCTURE CONCRETE ESTIMATE. The estimate sheet for the concrete is shown in Figure 4.12, and the unit costs for doing the work are shown in Table 4.4. These unit costs are indexed to the wage and rental rates listed at the top of the estimate sheet, and may be converted for use with other rates as discussed in Chapter 2.

Sections S_3, S_4, and S_5 are priced, and the time required for completion of the item is determined in the days column. In this case, we assume a four-carpenter crew for the foundations and for the headwalls. One day is added to the foundation time for the pour. We assume a six-carpenter crew for the remaining work, and add 8 working days to the time required to set and strip the walls and top slab. This allows 1 working day to make the last pour, 5 working days to cure and attain a concrete strength of 1500 PSI, and 2 working days to complete the finishing work. All the work is on the critical path for the item.

With the time to complete the item established, the general cost section is completed and the bare cost of the item is obtained. The allowance for payroll expenses is determined by the methods set forth in Chapter 3 for this type of work.

4.3.2 Remove Bridge

We will assume that the bridge to be removed is a typical old timber bridge, spanning the channel indicated by the OG in Figure 4.9. During the job site inspection, the bridge is noted to be as follows:

Deck	28 feet wide by 47 feet long. Height to top of deck 8 feet. Deck is made of laminated 4 × 6s and is surfaced with 1½ inches of cracked plant-mix surfacing.
Spans	3 at 15.33 feet each.
Stringers	6 × 16s–18 feet long, 14 inches center to center.
Caps	4 each 12 × 14–32 feet long.
Abutments	7 piles each. Piles are 12″ in diameter. 4 × 12 sheeting 5 feet vertical by 46 feet wide each abutment.
Bents	5 piles each. Piles are 12″ in diameter. 4 pcs. 3 × 8 v bracing by 18 feet long each bent.
Railing	8 × 8 posts. 18 pcs. 6 feet long. 6 × 6 rail. 256 linear feet. 8 × 8 curb. 96 linear feet.

ITEM 3. 150 CY STRUCTURE CONCRETE (153.0)

	C 16.34 K_{6+2} 1000 L 13.48 F 16.10 K_{cr} 75			L	DAYS	M,R
	S_0 GENERAL COSTS			}		}
	CONC. 153.0×1.02 = 156 CY		30			4680
	8.1 MFBM BUY		300			2430
	0.7 ✓ HAVE		100			70
	1.5 MSF BUY		500			750
	4.0 ✓ HAVE		170			680
	TIES 4.0C 1 MO.		40/C			160
	MINOR EXP. 0.153 MCY		3450			530
	MAKE FORMS 3282 SF		0.23	755		—
	ADD'L. FINISH 0.153 MCY		3300	505		20
	✓ CURE 3 WKS.		50	150		—
	SAND BLAST 0.153 MCY		550	85		15
	COMPRESSOR 0.5 MO.		400			200
	VIBRATORS 0.5 MO.		200			100
	GENERATOR 0.5 MO		250			125
	DEVELOP WATER 50+100					150
	EJ MAT'L. 48 SF (½")		0.50			25
	BURLAP 1.6 MSF					50
	CLEAN UP 14.2 MFBM	43	12	610		170
	S_3 F'ND'S & BOT. SLAB			}		}
	8CY ADD'L. CONC.	2	34	15	1770	270
	2023 SF FG		0.18	365	1000×⅔	
	2CY H EX		18	35	=3,+1	
	SS 177 SF H1½		2.70	480	=4D	
	SS 70 SF H0⁶		3.45	240		
	SS 265 SF H4		2.40	635		
60.0	POUR 61 CY	4.93	4.30	300	—	260
	S_4 HEADWALLS				1570	
	SS 1369 SF H12		1.15	1570	1000×⅔	
23.9	POUR 25 CY	6.88	6.64	170	=3D	165
	S_5 WALLS & TOP SLAB			—		
	SS 2958 SF H6		0.73	2160	3980	
	SS 243 SF H2		1.28	310	1000	
	SS 1440 SF SOFFIT		1.05	1510	=4,+8	
69.1	POUR 71 CY	5.21	4.96	370	=12D	350
153.0	Σ= $ 24700 ($161/CY) INCL. P.R. EXP. (1.315)			10265 13500	Σ 19D	11200

Figure 4.12 Estimate for box culvert concrete.

Table 4.4. Unit Costs for Box Culverts

Activity	Labor		Rental		Unit of Measure
	Unit Cost	Index	Unit Cost	Index	
General costs					
Additional finishing	$3300	$K_{L+F} = 29.58$	—		MCY
Sandblast	$ 550	$K_L = 13.48$	—		MCY
Foundations and bottom slab					
1. Fine grade	$ 0.18	$K_L \quad = 13.48$	—		SF
2. Hand excavation	$17.50	$K_L \quad = 13.48$	—		CY
3. SS headwall foundations	$ 2.70	$K_{6+2} = 1000$	—		SF
4. SS slab sides	$ 3.45	$K_{6+2} = 1000$	—		SF
5. SS cut-off walls	$ 2.40	$K_{6+2} = 1000$	—		SF
6. Or SS items $3+4+5$	$ 2.70	$K_{6+2} = 1000$	—		SF
7. Pour items $3+4+5$	$ 4.93	$K_{L+F} = 29.58$	$4.30	$K_{Cr} = 75$	CY
or	$ 9.67	$K_{L+F} = 29.58$	—		CY
Walls and top slab					
8. Make wall form panels	$ 0.23	$K_{6+2} = 1000$	—		SF
9. SS box walls	$ 0.73	$K_{6+2} = 1000$	—		SF
10. SS head walls	$ 1.15	$K_{6+2} = 1000$	—		SF
11. SS parapet walls	$ 1.28	$K_{6+2} = 1000$	—		SF
12. SS soffit forms	$ 1.05	$K_{6+2} = 1000$	—		SF
13. Or SS items $9–12$	$ 0.91	$K_{6+2} = 1000$	—		SF
14. Pour items $9–12$ or $9+11+12$	$ 5.21	$K_{L+F} = 29.58$	$4.96	$K_{Cr} = 75$	CY
15. Pour head walls only	$ 6.88	$K_{L+F} = 29.58$	$6.64	$K_{Cr} = 75$	CY
or	$ 7.37	$K_{L+F} = 29.58$	—		CY

Note: The unit costs for pouring the top slab assume it will be covered with earth or surfacing. When pours are large or average in size, the unit costs for the crane include 1 hour of travel time. On small pours, travel time will be additional.

Based on removing the piles to a depth of 3 feet below grade, the total timber in the bridge is calculated to be 25.9 MFBM. It is assumed that the 11 tons of plant mixed surfacing will be left on the site and will be incorporated into the fill on the subsequent paving contract. It is also assumed that local ordinances prohibit burning, and that the material removed will be disposed of at a public dump located 12 miles from the job along typical country roads.

The estimate sheet for the bridge removal is shown in Figure 4.13. The unit costs for removing the timber bridge are shown on the estimate sheet, and are based on K_L = 13.48 and K_{Cr} = 75. These costs are applicable for bridges such as the one described, having deck areas ranging from 600 to 3500 SF.

The calculations required to determine the costs to load and dispose of the bridge material are shown at the bottom of the sheet. A 55-minute hour is used to determine loading and hauling times, and the 40 MPH average haul speed is a typical value for good country roads.

If considerable loading and hauling were involved, three trucks would have been used to keep the loader and crew working steadily. In this case, where only a half day is required, two trucks are used, and the loader and crew will collect material to be loaded while waiting. The calculation of the haul cost assumes that the superintendent will order the trucks to arrive 30 minutes apart.

The quantity of torn out lumber assumed per load is small and the loading time is slow, since the material will be broken and twisted and some pieces may still be bolted together. A small rubber-tired loader using its backhoe attachment as a boom is used to load out the material. It is relatively cheap, is readily available, and will be able to move in and out of the channel easily.

The time required to complete the item is determined in the days column, and amounts to three working days.

4.3.3   Structure Excavation

The estimate sheet for the structure excavation is shown in Figure 4.14. The unit costs used are based on K_L = 13.48 and K_{Cr} = 75, and are applicable to box culverts similar to the one being bid. The labor shown is for hand work, and if a grade checker were to be used 20 hours of grade checker time would be added. In our estimate we assume the superintendent will check the grade.

It should be noted that the structure excavation is listed as Item 2F on the engineer's estimate. The F designates that the item is fixed and that, barring a field change in the work required, the final pay quantity will be

ITEM 1 REMOVE BRIDGE

L 13.48 K cr 75 16 CY SEMI 30 ½ CY RT LOADER 25		L	DAYS	M,R
REM. BR. 1316 SF	0.60	0.77	790	$\dfrac{1010}{75\times8}$ =2D	1010
(OR 25.9 MFBM	30.50	39)			—
LOADER ON-OFF 2 HR		25			50
2 TRUCKS ON-OFF 2HR		30			60
LOAD 25,9 MFBM	7.03	4.34	180	1 D	115
HAUL ✓ ✓		10.10			260
DUMP FEE 7 LDS		7			50
INCL. PAY ROLL EXP. (1.315)			970 1275	3D	1545

$$\Sigma = ^\$ 2820$$

SEMIS 16 CY x ¾ EFFECTIVE = 12 CY/LD = 3.9 MFBM/LD

25.9/3.9 = 6.6 LDS.   SAY 7 LDS @ 3.7 MFBM

½ CY LOADER 0.288 MFBM IN 2 MIN.

55/2 X 0.288 = 7.9 MFBM/HR

LOADING TIME 3.9/7.9 X 60   30 MIN.⎫

TRUCKS ROUND TRIP 24/40 X 60 X 60/55   40   ⎬ 75

                 AT DUMP   5   ⎭

TRUCK CYCLE = 1.25 HRS.

∴ USE 2 TRUCKS   0.625 HRS/LD.

LOADER TIME   6 LDS @ 0.63 + 0.5 = 4.3   SAY 4½ HRS

LOADER COST   4.5 x 25/25.9 = $ 4.34/MFBM

LABOR COST   4.5 x 13.48 x 3/25.9 = $ 7.03/MFBM

TRUCK COST   1.25 x 30/3.7 = $ 10.14/MFBM

Figure 4.13 Estimate for remove bridge.

ITEM 2F 230 CY STR. EX.				

L 13.48 K$_{Cr}$ 75		L	DAYS	M,R
612 CY 0.40 2.65		245	$\dfrac{1620}{75 \times 8}$	1620
INCL. PAY ROLL EXP. (1.14)		245 280	= 3 D.	1620
$\Sigma = ^\$1900$				

DIRECT JOB OVERHEAD				
BONDS 47 M	7.50	—		350
SUP'T. 6 WKS	550	3300		—
RADIO 1.5 MOS	70			105
TOILET 1.5 MOS	60			—
ICE – CUPS				50
VEHICLES (2) 3 V-MOS	280			840
INCL. PAY ROLL EXP. (1.355)		3300 4470		1435
$\Sigma = ^\$5905$				

TIME EST.	CONSEC.	CONC'T.
MOVE ON-OFF	2	2
REM. BR.	3	2
STR. EX.	3	3
STR. CONC.	19	19
REINF.	6	3
	33	29 D

Figure 1.14 Estimate for structure excavation and direct job overhead, box culvert.

230 CY. This being the case, it is not necessary to check the engineer's estimate for the excavation yardage. Our figure of 612 CY represents the true CY of structure excavation based on 1:1 side slopes.

The time required to complete the item is based on the rental cost, and amounts to 3 working days.

4.3.4 Direct Job Overhead

The direct job overhead estimate is also shown in Figure 4.14. The items listed are taken from the direct job overhead check list presented in Chapter 3. The cost of the bonds is based on an approximate estimate of $47,000 for the total bid price. The allowance for payroll expenses is determined by the procedures set forth in Chapter 3.

The time required to complete the job is determined at the bottom of the direct job overhead sheet. The first column contains the times needed to complete the various items of work, as determined on the estimate sheets, and 33 working days would be required if all items were done consecutively. The second column contains the critical path time for each item, assuming the work to be overlapped as much as possible; the estimated time for completion is 29 working days.

4.3.5 The Estimate Summary Sheet

The summary sheet is shown in Figure 4.15. With the final cost determined, the allowance used for contract bonds on the direct job overhead sheet is reviewed, and we note that our allowance is $15 high. No change is made in the cost column because the amount is considered to be insignificant.

The bid unit prices are assigned in accordance with the rationale discussed previously in connection with the estimate for the retaining walls. Assuming identical cost columns, no two estimators would assign identical bid unit prices to all of the items. In no case, however, will an experienced estimator assign direct job overhead or his markup to items that may underrun the engineer's estimate upon completion of the work.

4.4 Falsework for Concrete Bridges

Unit costs of setting and stripping bridge falsework are shown in Tables 4.5 and 4.6, and in Figure 4.16. The use of this data is illustrated in the estimates for the various types of bridges that follow.

SUMMARY. BOX CULVERT					
			COST	UNIT PRICE	TOTAL PRICE
1	REMOVE BRIDGE	LS	2820	5000	5000
2F	STRUCTURE EX.	230 CY	1900	12	2760
3	STRUCTURE CONC.	150 CY	24700	2.11	31650
4	REINF. STEEL	22100 LB	5350	0.24	5304
	SMITH 24.20/CWT				
	DIRECT JOB OVERHEAD		5900		
		TOTAL	40670		
		+10%	4070		
		TOTAL	44740		44714

Figure 4.15 Estimate summary sheet for box culvert.

Table 4.5. Unit Costs to Set and Strip Bridge Falsework Consisting of Timber on Shore "X"[a] Towers.

Number of Shoring Units (in height)	Steel Shoring Towers $/CWT		Top Lumber[b] $/MFBM	
	Labor $K_{6+2} = 1000$	Rental $K_{Cr} = 75$	Labor $K_{6+2} = 1000$	Rental $K_{Cr} = 75$
1	$10.10	—	$148	—
3	$ 8.90	$0.60	$211	$54
5	$11.50[c]	$0.60	$300	$54

[a]The trademark Shore "X" is registered by Waco Scaffold & Shoring Co.
[b]Costs are based on using 4 × 12 or 6 × 12 stringers with a maximum length of 12 feet. Transverse beams range from 4 × 4 to 8 × 8 inches in cross section.
[c]Towers 4 or more frames in height require 2 × 6 cross bracing. Allow $745 per MFBM ($K_{6+2} = 1000$) for setting and stripping the 2 × 6 bracing.
Note: The unit costs shown include the costs to set and strip the footing pads for the towers, and include 1 hour of travel time for the crane. Increase the unit costs for labor by 65% when small amounts of falsework are used to form isolated bent caps. On such work, the on-and-off charges for the crane and the shoring must be added.

In all cases the cost to set and strip soffit form plywood located directly on the falsework is included in the unit costs to set and strip the falsework, and no additional allowance for this work is necessary.

4.4.1 Shore "X" Steel Shoring

Shore "X" is a registered trademark. The shoring is manufactured by Waco Scaffold and Shoring Co., and is widely used as falsework for reinforced concrete bridges and other heavy structures. A Shore "X" tower is shown in Figure 4.1.

The basic shoring unit consists of two welded X-braced end or base frames, 4 feet wide by 6 feet high, tied together with removable X braces. When assembled, the 6-foot-high unit is 4 feet wide by 4 or 10 feet long in plan, depending on the X braces used.

The two base frame legs are made of 2.375 inch OD steel tubing. As many as 9 base frame units may be stacked, allowing base frame tower heights to vary from 6 feet to 54 feet, in 6 foot increments.

Extension end frames are often required for the top tower unit. The legs of these frames are 1.90 inches OD, and telescope into the legs of the base frames below. The two legs of the extension frames are connected with

Table 4.6. Miscellaneous Falsework Costs

Activity	Labor		Rental		Unit of Measure
	Unit Cost	Index	Unit Cost	Index	
SS wide flange beams[a]					
16 WF 36–24'	$ 40	$K_{6+2} = 1000$	$29	$K_{Cr} = 75$	Each
24 WF 76–35'	$ 59	$K_{6+2} = 1000$	$55	$K_{Cr} = 75$	Each
Footing pads					
Make	$115	$K_{6+2} = 1000$	—		MFBM
SS flat pads 2' × 2'[b]	$ 7	$K_{6+2} = 1000$	—		Each
Hand excavation for pads on abutment slopes	$ 21	$K_L = 13.48$	—		CY
Hand backfill for pads on abutment slopes	$ 24	$K_L = 13.48$	—		CY

[a]Labor costs include the cost to set and strip lateral bracing for top flanges of beams. Rental costs assume easy access for crane.
[b]Pads set on flat surface.

Note: When 20 or more beams will be set or stripped at one time, the unit costs for the crane will include 1 hour of travel time. Unit costs to set and strip wood falsework are shown on Figure 4.16.

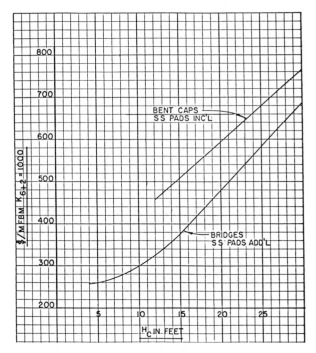

Figure 4.16 Unit labor costs to set and strip wood falsework. Posts and caps 8×8 maximum, stringers 6×12 twelve-feet long maximum. The cost to set and strip plyform that is directly on the falsework is included.

welded horizontal members located in the top foot of the 6 foot high frames, and are supported at the bottom by high strength steel pins which pass through the legs of the base frames below. The pin holes are 1 foot apart vertically, and thus permit the extension frames to extend from 1 to 5 feet above the base frame units. The extension frames are braced with removable X braces in both directions.

Removable screw jacks with 8-inch-square base plates are used at the top and bottom of each tower leg for grade adjustment, and extend from 4 to 12 inches. The bottom jacks are normally set on 2-foot-square wooden footing pads. A 3 × 8 rough cap 6 feet long is permanently attached to the top jacks at each end of the tower. The costs to set and strip the typical footing pads and these caps are included in the unit costs to set and strip steel shoring towers.

A maximum of two additional stacks of end frames may be attached to one or both ends of a shoring tower. Such additional frames increase the plan length of the tower by 8 inches per additional stack of frames. In

Table 4.7. Shore "X"¹ Tower Weights

Tower Description		Base Frame 65#	Ext. Frame 50#	Adapter Pin 1#	Coupling Pin 2#	End Cross 11#	Side Cross 18#	Screw Jack 19#	Screw Jack 20#	Timber Clamp 2#	Scraper Bar 1#	Wt.
2 Frame tower												
1 BF	6'8" to 8'8"	2	—	—	—	—	2	—	8	—	—	326#
1 BF & E	7'8" to 13'8"	2	2	4	—	2	4	4	4	—	—	484
2 BF	12'8" to 14'8"	4	—	—	4	—	4	—	8	—	—	500
2 BF & E	13'8" to 19'8"	4	2	4	4	2	6	4	4	—	—	658
3 BF	18'8" to 20'8"	6	—	—	8	—	6	—	8	—	—	674
3 BF & E	19'8" to 25'8"	6	2	4	8	2	8	4	4	4	—	840
4 BF	24'8" to 26'8"	8	—	—	12	—	8	—	8	4	—	856
4 BF & E	25'8" to 31'8"	8	2	4	12	2	10	4	4	8	—	1022
Extra BF	6'	2	—	—	4	—	2	—	—	—	—	174

3 Frame tower

											Weight
1 BF	3	—	—	—	4	—	12	—	—	—	507#
1 BF & E	3	3	6	—	8	3	6	6	—	—	762
2 BF	6	—	—	6	8	—	12	—	—	—	786
2 BF & E	6	3	6	6	12	3	6	6	—	—	1041
3 BF	9	—	—	12	12	—	12	—	—	—	1065
3 BF & E	9	3	6	12	16	3	6	6	6	—	1332
4 BF	12	—	—	18	16	—	12	—	6	—	1356
4 BF & E	12	3	6	18	20	3	6	6	12	—	1623
Extra BF	3	—	—	6	4	—	6	—	—	—	279

Doubled frame additive

											Weight
1 BF	1	—	—	—	—	—	4	—	4	—	149#
1 BF & E	1	1	2	—	—	2	2	2	8	—	214
2 BF	2	—	—	2	—	—	4	—	8	—	222
2 BF & E	2	1	2	2	—	2	2	2	12	—	287
3 BF	3	—	—	4	—	—	4	—	12	—	295
3 BF & E	3	1	2	4	—	2	2	2	16	—	360
4 BF	4	—	—	6	—	—	4	—	16	—	368
4 BF & E	4	1	2	6	—	2	2	2	20	—	433
Extra BF	1	—	—	2	—	—	2	—	4	—	73

Note: Weights shown are for towers 10 feet in length. When towers are 4 feet long, deduct 7 pounds for each side cross.

'The trademark Shore ''X'' is registered by Waco Scaffold & Shoring Company.

normal use, the towers are erected in rows parallel to the bridge center line. The spacing between the rows usually depends on the dimensions of the structure being supported, and the spacing between the towers in the rows is adjusted to bring the actual leg loads as close to the allowable working loads as possible. To attain this goal, it is sometimes necessary to erect three-frame towers. Such towers are made by attaching an additional stack of end frames to the typical tower units with X braces. Thus a three-frame tower may be 4 × 8 feet, 4 × 14 feet, or 4 × 20 feet in plan.

The allowable working loads for steel shoring are based on tests by the manufacturer and on regulations set by governmental agencies. In our designs, we will assume a working load of 10 kips per leg. Thus a typical four-legged tower will support a total load of 40 kips, and a tower with doubled end frames at each end will support a total load of 80 kips.

Loading only one leg of a frame or one frame of a tower can cause failure. In our designs, we will distribute the loads equally to the two legs of each end frame, and will not exceed 10 kips of differential loading between the frames of any tower.

Since shoring towers are erected in various configurations, the unit costs to set and strip this material are based on the CWT of steel involved for towers of a given number of units in height. Table 4.7 gives the tower weights for Shore "X" steel shoring, and our estimates will be based on these weights. A tower of two base frames plus an extension frame in height is designated "2B + E" and is considered to be a three-unit tower in assigning the unit costs shown in Table 4.5.

4.4.2 Heavy Timber Bents

Heavy timber bents supporting wide flange beams may be required when roadways must pass through bridge falsework. The costs to fabricate, erect and dismantle such bents are discussed in Chapter 6.

4.4.2.1. SAND JACKS. Sand jacks are often used in lieu of wedges to facilitate the release of heavy timber bents. In such cases, the bent posts are set on, and are supported by, a 3- to 4-inch deep bed of confined sand. During the stripping operation, the sand is freed and removed to release the bent.

The sand may be confined by three layers of 2 × 4 material laid flat in the form of a square ring with clear inside dimensions at least 6 inches greater than the post dimensions. A close-fitting bearing plate is placed over the sand to prevent it from boiling upward as the bent is loaded. In most cases the compressive stresses in the posts are small, and a piece of ⅝-inch plywood will be used for the plate.

4.5 Driven Piles

Pile driving requires highly specialized equipment, crews, and experience. On all building work and on most engineering work driven piles are furnished and installed by pile driving contractors on a subcontract basis. In our estimates we will assume all driven piles to be furnished and installed by subcontractors.

Many pile driving contractors perform as general contractors on jobs where pile driving, heavy timber work, or underwater work constitute a major portion of the work to be done. This is commonly the case on waterfront work and on bridges where wide bodies of water are involved.

The general contractor will sometimes drive wood piles with his own forces when a few piles are required for temporary structures at a remote location. In such cases, the work is done with a truck crane equipped with a set of hanging leads and a drop hammer. A 3-man, inexperienced crew will drive 30-foot piles to a penetration of 20 feet in sandy loam soil at a rate of 8 to 10 piles per 8-hour shift.

Indexed to $K_{6+2} = 1000$, the piles will be cut off to grade at a cost of about \$7 per cut. In addition, about 6 hours will be required to rig up and dismantle the leads and hammer.

4.5.1 Pile Extensions

When concrete pile bents are required for structures built on dry land, the top of the driven portion of the piles is often set at 1 foot below the finished ground surface. The exposed portions of the piles are called pile extensions, and are formed and poured as columns.

The driven portion in such cases usually consists of a steel shell or pipe, filled with concrete. The combination of such piles and their extensions costs less than full length precast concrete piles. In addition, the use of pile extensions enhances the appearance of the exposed bent, since perfect alignment of the extensions can usually be achieved.

Pile extensions are usually constructed by the general contractor, and unit costs for this work are presented with the estimate for the flat slab bridge in section 4.7.

4.6 Cast-in-Drilled-Hole Concrete Piles

CIDH piles are frequently used in bridge construction when soil conditions are favorable and when the water table is known to be below the tip elevation of the piles. When holes must be drilled below the water table,

Table 4.8. Unit Costs for Cast-In-Drilled-Hole Piles and Pile Extensions

Activity	Labor		Rental		Unit of Measure
	Unit Cost	Index	Unit Cost	Index	
16"ϕ CIDH piles					
Pour directly from truck	$ 6.00	$K_L = 13.48$	—		CY
Drill, set reinf. steel, and pour	$ 0.60	$K_L = 13.48$	$ 1.46	$K_{BD} = 39^a$	LF
16"ϕ Pile extensions					
SS sonotube, set reinf. steel, and pour	$ 5.65	$K_{6+2} = 1000$	—		LF
Or:					
SS sonotube	$ 3.80	$K_{6+2} = 1000$	—		LF
Set reinf. steel cages	$ 0.29	$K_L = 13.48$	—		LF^b
Pour from deck forms^c	$ 1.55	$K_L = 13.48$	—		LF
Pour with crane	$12.00	$K_L = 13.48$	$28.00	$K_{Cr} = 75$	CY

$^a K_{BD}$ = hourly rental rate for truck-mounted, rotary bucket drill rig, operated and maintained. On-and-off charges are additional.
^{b} Per linear foot of cage.
^{c} Includes the costs to buggy concrete for bridges about 100 feet in length.
Note: Drilling and pouring costs, based on linear feet, are directly proportional to the square of the hole diameter. When 11 CY or more of pile extension are poured at one time, the unit cost for the crane will include 1 hour of travel time.

casing and pumps must be used, and the cost advantage of CIDH piles over driven piles is lost. In rare cases, the underground water may be under pressure. The drilled holes then act as artesian wells, and the installation of CIDH piles in such circumstances is virtually impossible.

When large amounts of CIDH piles are required, the work is normally done by drilling subcontractors using crane-mounted auger-type drills. Where small quantities are involved, at locations far from drilling contractors, the drilling can be done with truck-mounted rotary bucket drill rigs, which are commonly available.

Unit costs for CIDH pile work based on using a bucket drill rig and dry holes are given in table 4.8, and their use is illustrated in the cost estimate for the flat slab bridge.

4.7   Flat Slab Bridges

Complete bids were prepared in connection with the retaining walls and box culvert discussed in the preceding sections. A cost estimate will now be prepared for the bridge shown in Figure 4.17. This bridge is assumed to be a portion of a larger project, and cost estimates will be illustrated for the concrete, the CIDH pile, and the metal beam bridge rail items. The items for structure excavation, structure backfill, and reinforcing steel will not be considered. Excavation and backfill are discussed in Chapter 8, and reinforcing steel is a subcontract item in all cases.

It is assumed that the job site inspection reveals conditions similar to those noted in the previous estimates, and that the water table is known to be below the proposed tip elevation for the piles.

4.7.1   Structure Concrete

It is assumed that the flat slab bridge constitutes the total concrete item for the job. This assumption is necessary because the items in the general costs section of the estimate for the concrete depend upon the total structure concrete involved in the project.

4.7.1.1 THE CONCRETE TAKE-OFF. The concrete take-off is shown in Figures 4.18 and 4.19, and includes a take-off of the falsework lumber required. The falsework lumber is based on a preliminary design, and is illustrated in Figure 4.20. In practice, the falsework lumber take-off would be based on sketches made on the design sheets. Our take-off assumes that the contractor has all of the falsework lumber except the 2×4s in his yard.

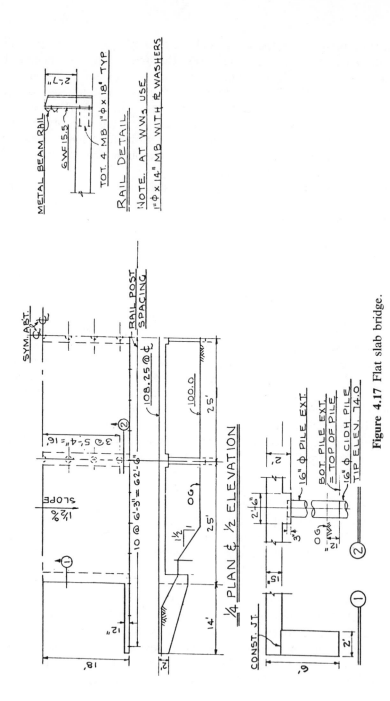

Figure 4.17 Flat slab bridge.

Figure 4D Flat Slab Bridge, Goose Lake Slough. Contractor: Tumblin Company.

The form lumber is taken off in the general costs section S_0, and is based on areas from sections S_6 and S_7. The falsework lumber is taken from section S_8. It is decided to form and pour the abutments one at a time. Forms are made for one abutment plus one additional back face, since the back face forms will be left in place until the deck slab is poured. An allowance is made for re-using the abutment forms in the deck formwork.

4.7.1.2 THE STRUCTURE CONCRETE ESTIMATE. The estimate sheet for the structure concrete is shown in Figure 4.21, and the unit costs for doing this work are shown in Tables 4.6 and 4.9, and in Figure 4-16.

The cost of minor expenses is determined by the procedure illustrated in Figure 4.5, since no recent cost data are available. The cost of 120 pairs of 2×4 wedges is included with the required chamfer strip under the millwork section. The allowance for the concrete buggy and collection hopper represents an expense that is actually related to the CIDH item, but is included here because it appears on the checklist and is in this case a very minor item of expense.

The time estimate is made in the days column, and a two-carpenter crew is assumed for the abutment work. 6 days are required to do the formwork, but only 3 days are on the critical path for the item. 3 additional days are added to pour, cure, and backfill the front face of the last abutment, and the estimated critical path time to complete the abutments is 6 days. A six-carpenter crew is assumed for the falsework, and 5 days

FLAT SLAB BRIDGE

			FG	HG	CY
ABUTM'TS & WINGWALLS					
2×4.75 × 36/27			72		12.7
34×6 + 36×4.75 + 2×2×4.75				394	
4̄ × 14 × ½7 × 2					4.2
(4̄ × 14 × 2 + 2×1 + 1.25×1) 2			56	231	
I ADD'L ABUTMENT COMPL.			128	625	16.9
		S₆	256	1250	33.8

	MISC.	CAP BMS.	DE	SOF	CY
DECK SLAB					
100 × 36 × 1.25/27					166.7
96 × 36				3456	
2×1.25×100 + 6×2.5×0.75			251		
3 × 2.5 × 0.75 × 36/27					7.5
6 × 0.75 × 36		162			
SET 42 SETS POST BOLTS	1				
S₇	1	162	251	3456	174.2

$$\Sigma_6^7 S = 208.0 \text{ CY}$$

		PADS	6×6 4×4,6	6×12	2×
FALSEWORK LUMBER H꜀7					
END BENTS 4×6 8/10		18	0.16		
TYP. BENT CAP 6×6 2/20			0.12		
POSTS 4×6 RO 9/6		9	0.11		
BRAC'G. 2×4 8/12					0.07
9 ADD'L BENTS		81	2.07		0.63
LONGIT. BRAC'G (5 LINES)					
2×4 30/8 70/12					0.72
STRINGERS					
6×12 72/12				5.2	
X BMS					
4×4 102/5			0.68		
2×3 BLOCK 102/1					0.05
4×4 68/10 × 2 SPANS		= 1.7	1.81		
4×4 72/10 X ✓		MFBM	1.92		
HAVE 13.8 MFBM S₈		108	6.9	5.2	1.5
BUY 1.5 ✓					

Figure 4.18 Concrete take-off for flat slab bridge.

FLAT SLAB BRIDGE (CONT'D)	MISC.	SF MAKE	MFBM HAVE	MSF PLY	2. MFBM 2X
<u>GENERAL COSTS</u>					
S_6 MAKE $1250 \times \frac{1}{2} + 6 \times 34$		829			
$\qquad 829 \times 1.17,\ 1.75$				1.0	1.4
$S_7\ \ 413 \times 1.17,\ 1.75$		251		0.5	0.7
$\qquad 3456 \times 1.17$				4.0	
RE-USE ABUTS					
$\qquad (829 - 2 \times 6 \times 34) \frac{2}{3} \times 1.17, 1.75$				(0.3)	(0.5)
S_8			13.8		1.5
FORM TIES $625/8 = 78$	1				
S_0	1	1080	13.8	5.2	3.1
				HAVE 4.0	

Figure 4.19 Concrete take-off for flat slab bridge (continued).

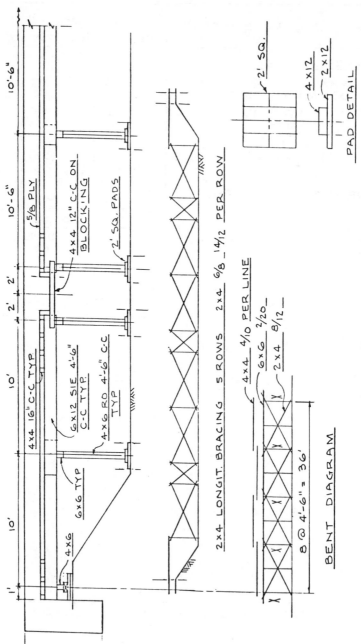

Figure 4.20 Falsework for flat slab bridge.

ITEM 3. 208 CY STRUCTURE CONC. (208.0)

C 16.34 K$_{6+2}$ 1000 L 13.48 F 16.10 K$_{cr}$ 75		L	DAYS	M,R
S$_0$ GENERAL COSTS				—
CONC, 208.0 ×1.02 = 212 CY	30			6360
3.1 MFBM BUY	300			930
13.8 ✓ HAVE	120			1655
1.2 MSF BUY	500			600
4.0 ✓ HAVE	170			680
TIES 0.8C 0.5 MO.	40/c			15
MINOR EXP. 0.208 MCY	3000			625
MAKE FORMS 1080 SF	0.23	250		—
ADD'L FIN. 0.208 MCY	2750	570		20
ADD'L CURE 3 WKS.	50	150		—
SAND BLAST 0.208 MCY	390	80		15
COMPRESSOR 0.5 MO.	400			200
VIBRATORS 0.5 MO.	200			100
FIN. MACHINE 3600 SF	0.05			180
GENERATOR 0.5 MO	250			125
DEVELOP WATER 50+100				150
BUGGY + HOPPER				30
CURE RUGS				100
SWEEP 3.6 MSF	13	45		—
CLEAN. UP 16.9 MF·BM 43	12	725		200
S$_6$ ABUTM'TS			1990	
FG 256 SF	0.40	105	1000×⅔	
SS 1250 SF H6	1.51	1885	= 6	
POUR 35 CY	4.35	150	3+3=6D	
S$_8$ FALSEWORK			4620	
MAKE PADS 1.7 MFBM	115	195	1000	
SS 108 PADS	7	755	= 5D	
SS 13.6 MFBM Hc7	270	3670		
S$_7$ DECK SLAB			1295	
SET 42 SETS AB	11	460	1000×⅔	
SS 162 SF BMS	0.87	140	= 2	
SS 251 SF DE	2.76	695	2+11=13D	
SS SOF. PLY INCL W. FWK.		—		
POUR 178 CY 8.54	4.40	1520		785
		11395		12770
$\Sigma = \$ 27755$ ($133/cy) INCL. PR EXP. (1.315)		14985	Σ 24D	

(left margin notes: 33.8 ; 174.2 ; 208.0)

Figure 4.21 Estimate for flat slab bridge concrete.

Table 4.9. Unit Costs for Flat Slab Bridges

	Labor		Rental		Unit of Measure
Activity	Unit Cost	Index	Unit Cost	Index	
General costs					
Make form panels[a]	$ 0.23	$K_{6+2} = 1000$	—		SF
Additional finishing	$2750	$K_{L+F} = 29.58$	—		MCY
Sandblast[b]	$ 390	$K_L = 13.48$	—		MCY
Abutments and wing walls					
Fine grade	$ 0.40	$K_L = 13.48$	—		SF
SS H 5–6	$ 1.51	$K_{6+2} = 1000$	—		SF
Pour	$ 4.35	$K_L = 13.48$	—		CY
Deck slab					
SS deck edge	$ 2.76	$K_{6+2} = 1000$	—		SF
SS pile cap side forms	$ 0.87	$K_{6+2} = 1000$	—		SF
Pour[c]	$ 8.54	$K_{L+F} = 29.58$	$4.40	$K_{Cr} = 75$	CY
Set rail post anchor bolts	$11.00	$K_{6+2} = 1000$	—		Set
Set posts and metal beam rail	$ 2.38	$K_{6+2} = 1000$	—		LF
SS formed deck drains	$ 8.15	$K_{6+2} = 1000$	—		Each

[a]Includes the costs to make abutment and deck edge forms.
[b]Includes the costs to sandblast the abutment construction joints and the tops of piles and pile extensions.
[c]If bridge deck finishing machine is not used, increase the unit cost for labor by $1.00 per CY.
Note: When pours are large or average in size, the unit costs for the crane include 1 hour of travel time. On small pours, travel time will be additional.

will be required. A four-carpenter crew is assumed for the deck slab work. 2 days are required, and a total of 11 days are added, allowing 1 day to pour the slab, 8 working days to cure prior to stripping the falsework, and 2 days to complete the additional finishing. The total time required to complete the item is 24 working days.

4.7.2 CIDH Piles

The estimate sheet for this work is shown in Figure 4.22, and the unit costs for doing this work are shown in Table 4.8.

The pay length for CIDH piles is measured from the top of the pile extension to the bottom or tip of the pile. Two types of work are involved, and the quantities are taken off and estimated separately as shown. This allows a quick comparison to be made on bid day, if a drilling subcontractor submits an unforeseen bid for the CIDH portion of the work.

The reinforcing steel for CIDH piles is not included in the job item for reinforcing steel, and its cost must be included in the cost of the pile item. The reinforcing steel for pile extensions normally consists of six number 6 vertical bars wrapped with W3.5 wire with a 6-inch pitch. The vertical bars extend 18 inches into the concrete above. The reinforcing steel for the CIDH portion normally consists of four number 6 bars, wrapped as before, with the vertical bars extending 18 inches into the pile extension. The reinforcing steel for the CIDH portion may or may not extend to the full depth of the pile. In our case, we assume these cages are specified to extend 12 feet below the joint and the total length of these cages will be 13.5 feet each.

Since the linear feet of required cages is never equal to the linear feet of piles plus extensions, it is important for the contractor to determine exactly what the steel subcontractor is quoting when the cages are bid on a per foot basis.

The cost of disposing of the material excavated from the drilled holes is not included in the unit costs shown in Table 4.8. In our case, we assume this material may be left in the channel bottom.

The time estimate for the item is made in the days column of the estimate sheet. The time to drill and pour the piles is determined by dividing the rental cost of the drill rig by the daily rental rate, and amounts to 3 days. The three days required to do the pile extensions are based on the total labor cost and an assumed two-carpenter crew.

4.7.3 Metal Beam Bridge Rail

The estimate for the bridge rail is also shown in Figure 4.22, and the unit prices are shown in Table 4.8. The required anchor bolts, posts, rails, end

ITEM 5. 683 LF CIDH PILES (682.5)			
L 13.48 K$_{6+2}$ 1000 C 16.34 F 16.10 BUCKET DRILL 38	L	DAYS	M, R

S$_1$ 525 LF CIDH
 	
CONC. 27.1×1.08=29.3 CY   30
REINF. CAGES 283.5 LF   2
DRILL, SET CAGES & POUR
  525 LF   0.60   1.46
B. DRILL ON-OFF 2 HRS.   38
 	
          INCL. PAY ROLL EXP. (1.305)
    Σ$_1$ =$2700   ($5.14/LF)
 	
S$_2$ 157.5 LF PILE EXTENSION
 	
CONC. 8.1 ×1.02 = 8.3 CY   30
REINF. CAGES 189 LF   2.80
168 LF 16"φ SONOTUBE   3.75
SS SONO., SET CAGES
  AND POUR 157.5 LF   5.65
 	
          INCL. PAY ROLL EXP. (1.315)
  Σ$_2$ =$2580   ($16.38/LF)
 	
  Σ$_1^2$ =$5280
	{ 315
	—
	315
	410
	
	{ 890
	890
	1170
		765
		8×38
		= 3D
		
		890
		1000×⅓
		= 3D
		
		Σ=6D
			{ 880
			565
			—
			765
			80
			2290
			
			250
			530
			630
			{
			{
			1410

ITEM 6. 260 LF BRIDGE RAIL (260)
 	
ALL MAT'L. INCL. 4 END SEC'S.
  AT 2.5' EA.   260 LF   16.50
SET 260 LF   2.38
 	
          INCL. PAY ROLL EXP. (1.315)
  Σ =$5105   ($19.63/LF)
	{ 620
	620
	815
		620
		1000×⅓
		= 2D
		
		Σ 2D
			—
			4290
			{
			4290

Figure 4.22 Estimate sheet for CIDH piles and metal beam bridge rail.

126

Figure 4E T-Girder Bridge with Pile Extensions. Kern River. Contractor: Tumblin Company.

sections, and rail bolts are normally furnished by a single material supplier. The cost to set the anchor bolts for the posts was included in the cost to form the bridge deck slab.

The time required to set the rail is determined in the days column, and a two-carpenter crew will require two days to do this work.

4.7.4   Total Time Required

The total time required to complete the flat slab bridge, including the CIDH piles and the bridge rail, will be 30 working days, assuming the structure excavation and backfill are done by others and are not included. The drilling will be done concurrently with the abutment work, the pile extensions will be done concurrently with the falsework, the railing will be set while the structure is curing, and the final cleanup can be done concurrently with the stripping of the falsework and the final additional finishing operation. 24 working days are required to set, strip, and pour the bridge. One day is added to place reinforcing steel in one abutment, three working days are added to place the reinforcing steel in the deck slab, and two working days are allowed for moving on and off the job and to cover a possible delay resulting from the tight scheduling of the abutment and CIDH pile work.

4.8   Bridge Bents and Columns

Steel girder bridges and prestressed concrete girder bridges are frequently supported on concrete bents such as the one shown in Figure 4.23.

Figure 4F Bridge Bent, Plate Girder Bridge. Contractor: Tumblin Company.

4.8.1 The Concrete Take-off

The concrete take-off for the bent is shown in Figure 4.24. The general costs section of the take-off is omitted, since it depends on the total quantity of structure concrete on the job.

The required timber falsework is shown in figure 4.25, and is based on a preliminary design. In practice, no formal falsework drawing would be required at the time the estimate was being prepared, and the falsework lumber take-off would be based on sketches from the design sheets.

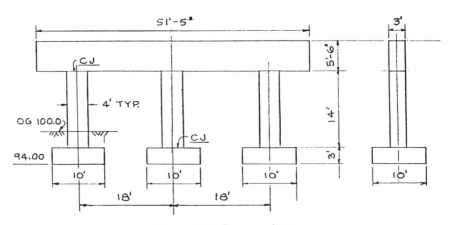

Figure 4.23 Concrete bent.

CONCRETE BENT

	MISC	FG	H3	CY
FOUNDATIONS				
$3(10 \times 10 \times {3}/{27})$		300	360	33.3
POUR NEAT $3(11 \times 11 \times {3}/{27}) - 33.3 = 7.0$ CY	I			
S_9	I	300	360	33.3
			H 14	CY
COLUMNS				
$3(4 \times 3 \times {14}/{27})$ S_{10}			588	18.7
		SOF	H5.5	CY
CAP				
$3 \times 5.5 \times 51.42/27$ S_{11}		119	599	31.4
$\Sigma_9^{11}\, S = 83.4$ CY				

	MISC	4×12	4×6 6×6	2×
FALSEWORK				
I BENT 6×6 $^1/5$, 4×6 RO $^2/12$			0.063	
2×4 $^2/5$ $^4/8$				0.028
5 BENTS ADD'L.			0.315	0.140
BRAC'G. BETW'N. BENTS				
2×4 $^8/12$ $^2/8$ $^{18}/6$				0.245
STRINGERS 4×12 $^4/14$ $^8/12$		0.608		
X-B'MS 4×6 $^{36}/5$ $^8/8$			0.488	
12 PR. 2×4 WEDGES } 4 PADS 2' SQ = 0.1 MFBM }	I			
TOTAL SS 1.9 MFBM H_C 14 S_{12} ALL WOOD	I	0.6	0.9	0.4
OR USE				
2 - $\curvearrowright$ 18 + E = 15.3 CWT				
PLUS 0.608 + 0.488 = 1.5 MFBM				

Figure 4.24 Take-off for concrete bent.

Steel shoring towers can be used to replace the posts, caps, and bracing shown in Figure 4.25, and if used will reduce the cost to set and strip the bents by a substantial amount. In this case, two 3-frame, 1 B+E towers would be required.

4.8.2   Unit Costs

The unit costs to do this work are shown in Figure 4.26 and Table 4.10. Unit costs shown in Figure 4.26 are applicable to the column forms and to the cap forms when the bent cap is separate from the girder system, as shown. On many T-girder and box girder bridges, the bent columns extend to the bottom of the girders, and the bent cap is replaced with a bent diaphragm which is formed and poured with the girders.

Table 4.10 gives unit costs for setting and stripping column footings poured in neat excavations, as well as unit costs for setting and stripping footing forms when required. When footings are poured against the earth, the costs given cover the setting and stripping of the column starter walls and the reinforcing steel templates. These costs are included in the costs of setting and stripping footing forms when side forms are used.

When foundation concrete is poured in neat excavations, considerable savings can be realized. The savings in formwork, excavation, and backfill costs will usually, more than offset the cost of the additional concrete required.

In many sandy soils it is not practical to pour footings neat, and it is usually impractical when job conditions require the excavation to remain open for an extended period before pouring the concrete, or when the footings are small and relatively deep. Bridge column footings are suited to the neat pouring method, since they are regular in plan and can be poured quickly. In addition, their size is such that even if the banks slough in after the concrete has set up there will still be ample room to set and strip the columns.

Separate unit costs are given in Table 4.10 for pouring the columns and the bent caps. The columns must be poured before placing the reinforcing steel in the bent cap, since the heavy cap reinforcing will prevent the use of elephant trunks during the column pour. Elephant trunks are normally specified to be used in all cases where the free fall exceeds 6 feet. Where the fall exceeds 6 feet and there is enough space to use elephant trunks, their use is mandatory to prevent segregation. Elephant trunks cannot be used when the clear distance between curtains of reinforcing steel is less than 11 or 12 inches. This situation occurs frequently on high wall pours. In such cases, segregation is prevented by the proper use of splash

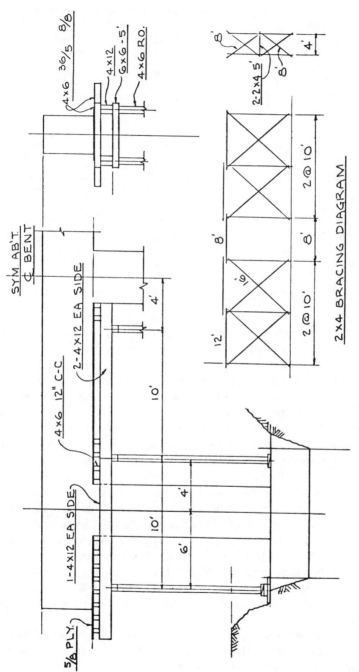

Figure 4.25 Falsework for concrete bent.

131

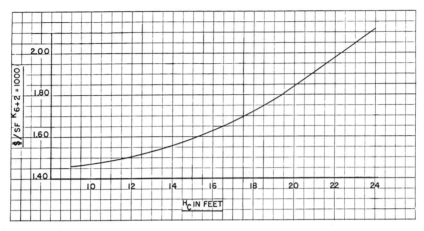

Figure 4.26 Unit labor costs to set and strip square or rectangular bent caps and columns.

boards, which, in conjunction with the wall forms themselves, will prevent segregation, provided the concrete is vibrated properly.

Wall pours as high as 28 feet are best poured by this method when the space available prevents the use of elephant trunks. Heavy-duty vibrators with long shafts are required, as are experienced vibrator men. Attempts to pour such walls through pour windows in the wall forms will necessitate more lateral movement of the concrete with the vibrators, are costly, and will result in poor work in nearly every case. The splashboards normally used are made of plyform, and are 22 inches wide by 8 feet long. They are suspended inside of the reinforcing steel curtains and are moved ahead as the pour progresses. Collection hoppers are used to direct the concrete between the boards.

Table 4.10 also gives unit costs for setting and stripping the round steel column forms frequently required in bridge construction. Although the unit costs for rental in connection with concrete work include a maximum of 1 hour on and off time for the crane, an exception is made in the case of setting and stripping steel column forms. For this work, the on and off charges for the crane must be added. This is necessary because only one or two sets of steel column forms may be available, and the number of units to which travel time would be distributed can easily vary by as much as 100 percent.

The costs shown for making, setting, and stripping wooden forms for round columns illustrate the advantages of using steel forms when large diameter round columns are required.

Table 4.10. Unit Costs for Concrete Bents and Columns

Activity	Labor		Rental		Unit of Measure
	Unit Cost	Index	Unit Cost	Index	
Foundations					
Fine grade	$ 0.28	$K_L = 13.48$	—		SF
SS (neat excavation)	$93	$K_{6+2} = 1000$	—		Each
or					
SS side forms	$ 1.80	$K_{6+2} = 1000$	—		SF
Pour foundations	$ 4.45	$K_L = 13.48$	—		CY
Bents—wood forms[a]					
Make form panels	$ 0.29	$K_{6+2} = 1000$	—		SF
Pour columns	$ 8.30	$K_L = 13.48$	$10.50	$K_{Cr} = 75$	CY
Pour caps	$ 5.60	$K_L = 13.48$	$ 7.80	$K_{Cr} = 75$	CY
Round columns—steel forms[b]					
SS	$360	$K_{6+2} = 1000$	$ 150	$K_{Cr} = 75$	Each
Pour	$ 8.30	$K_L = 13.48$	$10.50	$K_{Cr} = 75$	CY
Round columns—wood forms[c]					
Make forms	$ 2.95	$K_{6+2} = 1000$	—		SF
SS	$ 3.90	$K_{6+2} = 1000$	—		SF

[a]Columns and caps square or rectangular in cross section.
[b]Diameter 3–6 feet. Height 24–30 feet. On-and-off charges for crane are additional.
[c]4 feet in diameter by 12 feet in height.

Note: When pours are large or average in size, the unit costs for the crane include 1 hour of travel time. On small pours, travel time will be additional. Travel time for the crane used to set and strip steel column forms should be added in all cases. Unit costs to set and strip wood forms for rectangular or square bent columns and caps are shown in Figure 4.26.

Figure 4G Bridge Bents, Steel Beam Bridges. Contractor: Tumblin Company.

4.9   Bridge Abutments and Wingwalls

Typical bridge abutments are illustrated in Figures 4.27 and 4.28. Bridge abutments are not designated by type numbers in practice, and the numbers shown are for this discussion only. Depending on the conditions, piles may be used with any of the abutments shown.

The type 1 abutment and wingwalls are similar to, but deeper than, the abutments used on the flat slab bridge previously discussed, and are normally used on T girder and box girder bridges. When such bridges are prestressed, a footing pad is used below the abutment wall shown. Expansion joint material is used at the construction joint between the pad and the wall to form a hinged joint, and the wing walls are poured after the bridge has been post tensioned.

The type 2 and type 3 abutments are normally used on prestressed concrete girder and steel plate girder bridges. When type 2 abutments are used, the main foundation pad may return at the wingwalls for a short distance, and the small footing at the end of the wing walls is eliminated. Type 4 wingwalls are usually used with type 3 abutments, and are often built on stepped foundations.

Unit costs for abutments are shown in Table 4.11. When stepped foundations are involved, the earth will slough in behind the steps, as was discussed in connection with the box culvert cutoff walls, and an allowance must be made for the estimated amounts of additional concrete and step forms required.

For estimating, the contact area of the step forms required is based on the height of the step at its front face and on the distance between the

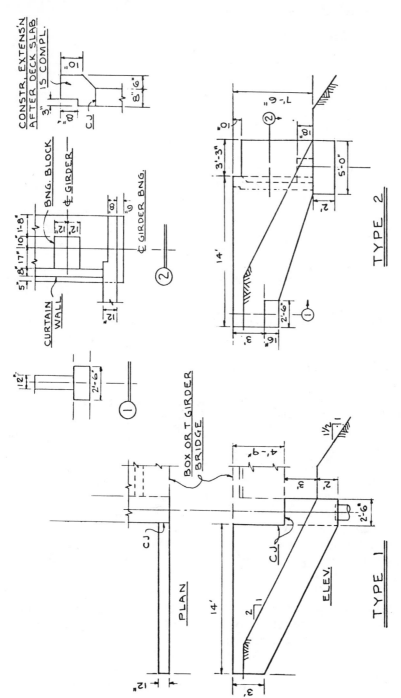

Figure 4.27 Bridge abutments.

135

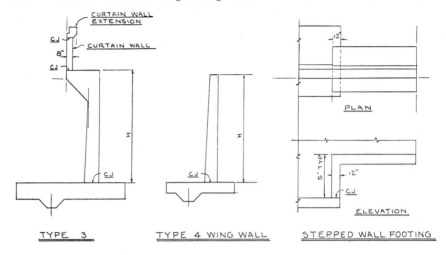

Figure 4.28 Bridge abutments.

foundation slabs at the sides. On type 2 abutments, the 8-inch walls at the sides are normally poured with the wingwalls, and are considered as wingwalls when applying the unit costs shown.

In designing the forms for type 3 abutments, the vertical haunch load due to the weight of the concrete and the live load is distributed to the back wall form studs, and to a beam and post arrangement at the rear edge of the haunch. The horizontal load from the fluid pressure of the concrete is held with form ties and wales that restrain the upper ends of the sloping haunch soffit joists. The unit costs shown for the setting and stripping type 3 abutments include the cost of setting and stripping the required beam and post system.

Small haunches involving increased wall thicknesses of 6 to 12 inches are sometimes required at the tops of walls. The forms for such haunches do not require the beam and post arrangement, and in such cases the unit price applicable to the wall without haunches is used to determine the setting and stripping cost, and the wall area used includes the actual contact area of the haunch form.

A comparison of the unit costs for retaining walls with those for type 4 abutment wing walls shows that the abutment work costs more, and that the unit costs realized for the type 4 wingwalls do not vary appreciably with the height of the walls normally encountered. This clearly illustrates the need to recognize the subtle differences inherent in seemingly similar types of work, to develop separate unit costs where required, and to use sound judgment in applying unit costs to proposed work.

Table 4.11. Unit Costs for Abutments and Wingwalls

Activity	Labor		Rental		Unit of Measure
	Unit Cost	Index	Unit Cost	Index	
Foundation slabs					
Fine grade	$ 0.28	$K_L = 13.48$	—		SF
Hand excavation	$14	$K_L = 13.48$	—		CY
SS H1.5–H2.5 level	$ 2.25	$K_{6+2} = 1000$	—		SF
SS H1.5–H2.5 stepped	$ 2.60	$K_{6+2} = 1000$	—		SF
SS H5 steps	$ 3.40	$K_{6+2} = 1000$	—		SF
Pour foundations	$ 5.00	$K_L = 13.48$	—		CY
Abutments and wingwalls					
Fine grade type 1	$ 0.40	$K_L = 13.48$	—		SF
Make form panels	$ 0.24	$K_{6+2} = 1000$	—		SF
SS type 1, complete	$ 1.95	$K_{6+2} = 1000$	—		SF
SS type 2 wingwalls	$ 1.95	$K_{6+2} = 1000$	—		SF
Pour type 1, complete	$ 5.25	$K_L = 13.48$	—		CY
Pour type 2 wingwalls	$ 5.25	$K_L = 13.48$	—		CY
SS type 3 H12–H24	$ 2.10	$K_{6+2} = 1000$	—		SF
SS type 4 H8–H18	$ 1.55	$K_{6+2} = 1000$	—		SF
Pour types 3 or 4	$ 5.85	$K_L = 1000$	$ 6.50	$K_{Cr} = 75$	CY
SS H4–H6 curtain walls	$ 1.75	$K_{6+2} = 1000$	—		SF
Pour curtain walls	$ 7.25	$K_L = 13.48$	—		CY
or	$ 6.50	$K_L = 13.48$	$11.00	$K_{Cr} = 75$	CY
SS bearing blocks	$ 1.95	$K_{6+2} = 1000$	—		SF
SS curtain wall extensions	$ 2.60	$K_{6+2} = 1000$	—		SF

Note: When pours are large or average in size, the unit costs for the crane include 1 hour of travel time. On small pours, travel time will be additional. Abutments and wingwalls, types 1–4, are shown on Figures 4.27 and 4.28.

137

Retaining wall foundations tend to be more regular in plan than abutment work, are simpler to lay out, and the steps, if any, are further apart. Larger areas of foundations and walls are normally involved in retaining wall work, and more efficient use of the crews is possible, since a daily and repetitive sequence of operations can be established.

4.10   Barrier Railing

Typical bridge barrier railings are shown in Figure 4.29. Although barrier rails are designated by type numbers in practice, the designations vary, and the type numbers shown here are for this discussion only.

Usually 100 linear feet of concrete barrier are set and poured each day. The front barrier rail forms are stripped and the exposed concrete is given a float finish on the same day the concrete is poured. Before acceptance, the back or outside face of the barrier and the concrete end posts, if any, are given a sacked finish.

Barrier railing is typically bid as a separate job item. In addition to the costs of setting, stripping, pouring, and finishing the concrete portion, the price bid includes the cost of the metal railing, which is furnished and set by a subcontractor. The cost of the reinforcing steel may be included in the price bid for the bridge reinforcing steel, or may be included in the

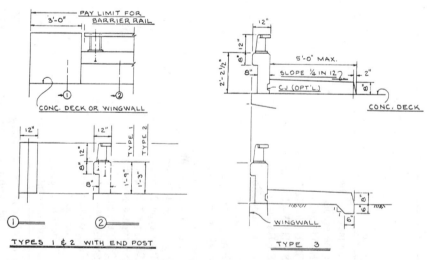

Figure 4.29 Concrete barrier railing.

Table 4.12. Unit Costs for Concrete Barrier Railing

Activity	Labor		Rental		Unit of Measure
	Unit Cost	Index	Unit Cost	Index	
Make wall forms	$ 1.17	$K_{6+2} = 1000$	—		SF
Make walk face	$ 0.25	$K_{6+2} = 1000$	—		SF
SS type 1	$ 1.60	$K_{6+2} = 1000$	—		SF
or	$ 6.13	$K_{6+2} = 1000$	—		LF
SS type 2	$ 6.13	$K_{6+2} = 1000$	—		LF
SS type 3 wall	$ 1.60	$K_{6+2} = 1000$	—		SF
SS type 3 walk face	$ 2.50	$K_{6+2} = 1000$	—		SF
SS type 3 wall plus walk, complete	$ 9.27	$K_{6+2} = 1000$	—		LF
Pour type 1, 2 or 3 walls[a]	$43.50	$K_{L+F} = 29.58$	—		CY
Pour type 3 walk	$26	$K_{L+F} = 29.58$	—		CY

[a]Includes the cost to float finish the front face of the wall. The cost to sack the back face, and the cost to sandblast the construction joint at the deck are additional.

Note: Barrier railings, types 1, 2, and 3, are shown on Figure 4.29.

price bid for the barrier railing, and the specifications must be checked regarding this in each instance. In all cases, the barrier reinforcing steel is furnished and placed by the reinforcing steel subcontractor doing the bridge reinforcing steel work.

Unit costs for barrier rails are shown in Table 4.12. The setting and stripping costs include the cost of setting the metal rail anchor bolts furnished by the railing subcontractor, as well as of setting expansion joint material, as required. The material cost of the anchor bolts is always included in the price bid by the steel subcontractor who furnishes and erects the steel on them. The unit costs for pouring include the cost of the float finish on the front face.

The setting and stripping costs shown are applicable to steel or wood forms when based on cost per linear foot of barrier. When wooden forms are used with type 3 barrier rail, the walk portion is poured separately, whereas when steel forms are used, the walk is poured integrally with the wall.

The pay limit for barrier railing is the measured length of the concrete barrier, and includes the length occupied by the end posts when concrete end posts are used. The additional formwork required for end post details is priced at the unit cost per square foot of contact area used for the barrier wall.

In practice, the barrier rail item is priced after the cost estimate for the bridges has been completed, since some of the general cost items for the barrier rail are related to those used for the bridge work. We will now present a complete cost estimate for a T-girder bridge with type 3 barrier rails.

4.11   T-Girder Bridges

The T-girder bridge is shown in Figure 4.30. Type 1 abutments with the dimensions shown in Figure 4.27 are used. It is assumed that the job conditions and location are the same as those assumed in the previous estimates.

4.11.1   Structure Concrete

It is assumed that the T-girder bridge constitutes the total structure concrete item for the project. We plan to pour the deck with the girders since considerable savings are realized by this method. The sandblasting otherwise required at the construction joint between the girders and the slab is eliminated, as is the cost of the after-pour cleanup and the additional curing expense required when the girders are poured separately.

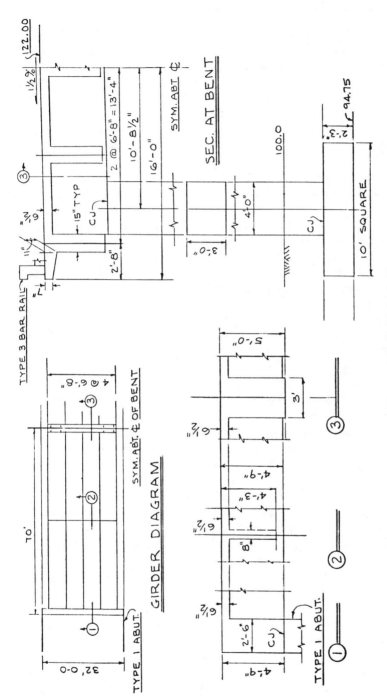

Figure 4.30 T-Girder bridge.

141

4.11.1.1 THE CONCRETE TAKE-OFF. The proposed falsework and slab soffit forms are shown in Figures 4.31 and 4.32, and are based on a preliminary design. The concrete take-off is shown in Figures 4.33 through 4.36, and includes the take-offs for the falsework and soffit forms.

As previously mentioned, the falsework take-off would normally be based on sketches made on the design sheets. It is of utmost importance, however, that the crews be furnished with good working drawings of the falsework before its erection. Good falsework drawings result in well-built falsework, and *vice versa*.

The limits for falsework and soffit formwork are shown in Figure 4.32. The girder form panels are shown as well, since the soffit form ledgers are nailed to the girder form studs. The labor cost to set and strip plywood that is directly on the falsework is included in the unit costs to set and strip the falsework.

The relationship between the three types of formwork shown in Figure 4.32 must be thoroughly understood. The girder form panels are made, set, and stripped at the unit costs for this work on a square foot basis; the slab soffit forms are set and stripped at their unit cost on a square foot basis, and the falsework is set and stripped at unit costs based on the MFBM of lumber and the number of beams or the CWT of steel shoring required. In all cases, the cost of the materials is based on the MFBM of lumber and the CWT of steel required.

The concrete take-off shown provides an orderly and systematic method for taking off and collecting the quantities required, and will now be discussed in detail.

The type 1 abutments and wingwalls are taken off first. The S_{13} total indicates that 216 SF must be fine graded, that 1436 SF of forms 5 to 10 feet in height must be set and stripped, and that 42.8 CY of concrete are required.

The S_{14} total indicates that 200 SF must be fine graded, that 180 SF of foundation forms 2.2 feet in height must be set and stripped, and that 16.7 CY of concrete are required. The miscellaneous item notes the additional concrete required if the footings are poured in neat excavations, in which case the 180 SF of forms will not be needed.

The S_{15} total indicates that 560 SF of column forms must be set and stripped, and that 17.8 CY of concrete are required.

The S_{16} total summarizes the deck and girder work, and shows the contact areas of the deck edge forms, girder soffits plus slab soffit, girder sides, and the amount of concrete involved. The approximate height of the deck edge forms and the girder side forms is indicated in the headings of these columns. The miscellaneous column notes the contact areas of deck and girder pour bulkheads since we plan to pour the combined deck

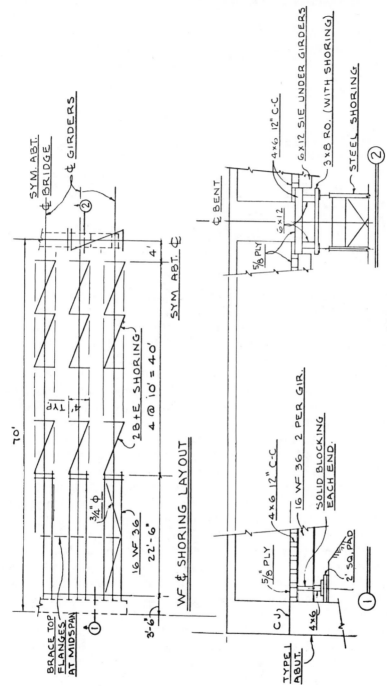

Figure 4.31 Falsework for T-girder bridge.

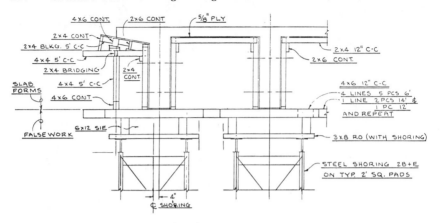

Figure 4.32 Falsework for T-girder bridge.

and girder system in two pours. The total areas of girder and slab soffits are also noted in the miscellaneous column.

The falsework take-off is summarized in the S_{17} totals, which indicate the CWT of 2 B + E steel shoring and the MFBM of lumber to be set and stripped, as well as the MFBM of 4 × 6 and 6 × 12 to be furnished. The miscellaneous column indicates the required number of flat pads for use under the WF beams and the number of WF beams required, and notes that the rod bracing for the top flanges of the beams will be taken off and priced on a separate work sheet.

The falsework take-off assumes the column footing excavations will be backfilled before erecting the falsework. This is good practice because important savings are realized when bridge falsework is erected on clean and approximately level surfaces. Excavation and backfill are separate job items, and are discussed in chapter 8. With proper scheduling, the backfilling can be done outside the critical path for the project.

In contrast, the falsework for the concrete bents discussed previously was erected before backfilling the column footing excavations. In that case, savings could be realized by using the cap falsework for staging during the column pours, and the construction sequence would be interrupted if the backfilling were done between the column and the bent cap operations.

The cost to set and strip the flat pads required for the steel shoring is included in the unit cost to set and strip the shoring, and we assume we have these pads in our yard with the shoring. The cost to set and strip the top flange bracing and the bridging at the ends of the beams is included in

T GIRDER BRIDGE		FG	H 5/10	CY
				I.
ABUTMENTS & WINGWALLS				
2.5 x 5 x 32/27		80		14.8
34.5 x 2 x 5			.345	
6.38 x 14 x 1/27 x 2		28		6.6
(6.38 x 14 x 2 + 3 + 4.75) 2			373	
ONE ADD'L. ABUTM'T.		108	718	21.4
S13		216	1436	42.8

	MISC.	FG	H 2½	CY
BENT F'ND'S.				
2(10x10x 2.25/27)		200	180	16.7
IF POUR NEAT ADD'L CONC.= 3.5 CY	1			
2(11x11x 2.25/27) -16.7				
S14	1	200	180	16.7

			H20	CY
BENT COL'S. 2(4x3x20/27) S15			560	17.8

	7"DE	SOF	H 4½	CY
GIRDERS + SLAB				
ABUT. DIAPHRAGM				
2.5 x 4.21 x 32/27				12.5
(4.75+4.21) 32 +2.5x4.75 x2- 4.21x1.25x5			284	
1 ADD'L DIA.			284	12.5
BENT DIAPHRAGM				
4.46x27.92-1.25x0.25x2 = 123.9 SF				
123.9 x 3/27				13.8
123.9x2+4.08x3x2 - 1.25 x 4.21 x10			220	
(27.92-8)3		60 G		
GIRDERS L=132 LF				
132 x 1.25 x 4.21/27 x5		825 G		128.6
(3.83+4.21)x132x2+4.21x2x132 x3			5457	
INTERMED. DIA. 5.42x3.71x0.67/27 x8			322	4.0
SLAB				
32 x140 x 0.542/27				89.9
(32-5x1.25)132+2.04x3x2		3411 S		
ADD'L. @ OH				
(0.75-0.54) 2.04x 135/27 x 2				4.3

Figure 4.33 Concrete take-off for T-girder bridge.

T GIRDER BRIDGE (CONT'D)	MISC.	7"DE	SOF	H4½	CY
GIR'S. + SLAB (CONT'D)					
DECK EDGE 135 × 0.583 × 2		158			
POUR BULKHEADS					
DECK 0.54 × 32 + 2.04 × 0.21 × 2 = 18 SF	1				
GIR'S. 5 × 1.25 × 4.21 = 26.5 SF					
GIR. SOF 885 SF ⎫ 4296 SF	1				
SLAB ✓ 3411 SF ⎭					
Σ_{13}^{16} S = 342.9 CY S₁₆	2	158	4296	6567	265.6

	MISC.	CWT 2B+E	MFBM SS	4×6	6×12
FALSEWORK					
BENT NI 2B+E 2 @ 9.45 CWT		18.9			
6×12 ⁶/₁₂ , 4×6 ²³/₆			0.71	0.28	0.43
2×4 FILL ⁴/₃₂			0.09	0.09	
GIR'S.					
NN 2B+E 10 @ 10.41 CWT		104.1			
NI 2B+E 10 @ 9.45 CWT		94.5			
6×12 ⁸⁰/₁₂ (SIE)			5.76*		5.76*
16WF36 ²⁰/₂₄ = 173 CWT	1				
STRUTS 4×6 60 LF			—	0.12	
RODS ON WORK SHEET	1				
END BENTS 4×6 68 LF			0.14	0.14	
FLAT PADS (4SF) 20 EA.	1		—	0.3	
4×6 X BEAMS					
STR. SPA 4' 3' 4' 2.7 4' 2.7 4' 3' 4					
L = 70 - 2.5 - 1.5 = 66' EA. SPAN					
67 + 1 @ DIA. = 68, × 2 = 136 LINES					
EVERY 5ᵀᴴ LINE CONT.					
136 × ⅕ = 28, + 2 + 2 = 32 CONT.					
4×6 ⁶⁴/₁₄ ³²/₁₂ ⁵²⁰/₆			8.80	8.80	
* SIE S₁₇	3	218	15.5	9.7	6.2

Figure 4.34 Concrete take-off for T-girder bridge (cont.).

| T GIRDER BRIDGE (CONT'D) | 4×4 | 3. |
	4×6	2×

SOFFIT FORM LUMBER

	4×6	2×
DECK OVERHANGS L=135' =27 @ 5' EA. SIDE		
SILL 4×6 $^{18}/_{16}$	0.58	
POSTS 4×4 $^{56}/_4$	0.30	
CAPS 4×4 $^{56}/_4$	0.30	
LEDGER +BRIDGING 2×4 $^{18}/_{16}$ $^{18}/_{16}$		0.38
BEAMS 4×6 $^{18}/_{16}$	0.58	
2×6 ✓		0.29
2×4 ✓		0.19
INTERIOR SOF. L=4 @ 32.67'		
1 BAY 2×4 $^{136}/_6$		0.55
2×6 $^{24}/_{12}$		0.29
3 ADD'L. BAYS		2.52
S$_{18}$	1.76	4.22

TOTAL SOF. $\dfrac{5.98}{3.41}$ = 1.75 $\dfrac{MFBM}{MSF}$

OH SOF. $\dfrac{2.62}{0.551}$ = 4.76   ✓

INT. SOF. $\dfrac{3.36}{2.86}$ = 1.17   ✓

Figure 4.35 Concrete take-off for T-girder bridge (cont.).

T GIRDER BRIDGE	MISC	MAKE	16WF36	2B+E	HAVE	PLY	2X	4.
GENERAL COSTS								
S_{13} ABUT'S. MAKE 1 718 × 1.17, 1.75		718				0.84	1.26	
S_{14} BENT FND'S. POUR NEAT 3.5 CY	1							
S_{15} COL'S. MAKE 2 560 × 1.17, 2.00		560				0.66	1.12	
S_{16} GIR'S + SLAB 6567 × 1.17, 1.75		6567				7.68	11.5	
4296 × 1.17						5.03		
158 × 1.17, 1.75		158				0.18	0.28	
RE-USE S_{13}, S_{15}								
PLY 0.84×2/3 } 0.66×1 }						(1.22)		
2× (1.26+1.12) 2/3							(1.59)	
S_{17} FALSEWORK WF BRACING	1		20 EA =17.3 CWT	218	15.9			
S_{18} SOF. FORMS					1.8		4.2	
TAPER TIES $\frac{6559}{8}$ = 820	1							
SHE-BOLTS (20-2)/2 = 9 SPA								
2 (10 ×5) = 100 BOLTS } 40 ½" φ × 46" } 60 ½" φ × 30 }	1							
S_0	4	8003	↑	218	17.7	(13.2)	16.8	
						HAVE 8.0		

Figure 4.36 Concrete take-off for T-girder bridge (cont.).

the unit costs to set and strip these beams. No allowance is made for buying the bridging material, since we plan to use scrap material for this.

A precise determination of the top flange bracing required to reduce the *ld/bt* ratio is not possible. When T-girder bridge decks and girders are poured together, the total dead load of the concrete must be taken by the falsework at one time and in view of the grave consequences of a falsework failure, we will use the rod and strut system indicated for estimation. Before erecting the falsework, we will investigate the bracing and beam system with respect to the horizontal loading involved.

SIE 6 × 12s are required for some of the falsework, but this poses no problem since wooden falsework beams are normally bought SIE. In this case the SIE beam is 9 percent stronger than an S4S beam, while the MFBM and the cost are the same for both.

The S_{18} total summarizes the soffit form lumber required, and is broken down into 4 × 4 and 4 × 6 lumber and 2-inch-thick material.

The conversion factors in MFBM per MSF for the soffit formwork are noted at the bottom of the sheet for rough checking of future take-offs. When slab soffit formwork constitutes only a small portion of the contractor's work, the material cost is often based on such factors, since the labor cost is based on the square feet involved and is not affected by inaccuracies in the conversion factors used.

No allowance is made on the take-off for the blocking or the short 2×4 post required at the deck overhangs. This material will be made from scrap lumber from the other formwork.

The general costs section is primarily a material summary, and is based on the S_{13} through S_{18} totals. The contact areas of the required forms are converted to MSF of plyform and MFBM of 2-inch-thick (2X) lumber using the factors discussed in the general costs section of this chapter. In this case, we plan to make one abutment form, and will re-use it on the second abutment. Two column forms will be made since the pours are small, only a small quantity of forms are involved, and the material will be reused in the soffit forms.

We assume we have used 4 × 4, 4 × 6, and 6 × 12 material in our yard, and the required MFBM of this material is summarized in the "have" column of section S_0. 13.2 MSF of plyform are required, and we note that we have in our yard, 8.0 MSF of used plyform that will be used for interior girder and soffit formwork.

The miscellaneous column notes that we will pour the column footings neat, and that 3.5 additional CY of concrete are required. The note to take off the wide flange bracing is made to ensure that this is not overlooked. The form ties needed are also taken off and noted in this column. 820 taper ties will be required for the girder form panels, and these same ties are ample for the abutment work.

Because of the lengths involved, she-bolts with ½ inch diameter inner tie rods will be used to tie the column forms. We will use three 36-inch and four 48-inch she-bolts for each 2 feet of height, and the bottom bolts will be located 1 foot above the bottom of the column form. 50 she-bolts are required per column, together with 20 tie rods 46 inches long and 30 tie rods 30 inches long.

The tie rod lengths are based on a rule of thumb that a 12-inch wall formed with 2 × 4 studs and double 2 × 4 wales will require a 2-inch tie rod. In this case, we estimate that 2 × 8 wales will be used with the 48-inch she-bolts, and that 2 × 6 wales will be used with the 36-inch she-bolts. When she-bolts are used, nearly all of the inner tie rods remain embedded in the concrete, and tie rods must be furnished for each use under normal circumstances. Although in our case no re-use is possible, the columns will be tied with two of the 36-inch bolts and both of the 48-inch bolts located just outside of the column form studs, and only one 30-inch tie rod would be embedded and lost for each set of 5 she-bolts used.

Typical she-bolts, taper ties, and snap ties are illustrated in Figure 4.37.

4.11.1.2   THE STRUCTURE CONCRETE ESTIMATE.   The estimate sheets for the structure concrete are shown in Figure 4.38 and 4.39. Work sheets relating to some of the general costs items are shown in Figures 4.40 and 4.41. Unit costs for the girder and deck work are shown in Table 4.13, and the unit costs for the abutments, columns, and falsework have been discussed previously.

The miscellaneous iron item includes the costs for the WF beam bracing, and column tie rods. The cost of a portable welding machine and a welder are included, since the bracing for the beams will be fabricated on the job.

The minor expenses item is set up and estimated on the work sheet, since this item consists primarily of material costs which cannot be indexed. Where recent cost histories for similar work are available, this item would be estimated on a cost per MCY of concrete involved.

The number of working days required to complete the work summarized in each section is based on using a four-carpenter crew on the abutments and columns, a two-carpenter crew on the column foundations, and six-carpenter crews on the girders plus slab and on the falsework. Two working days are added to the girder plus slab work to cover the two pours required, and eight working days are added to comply

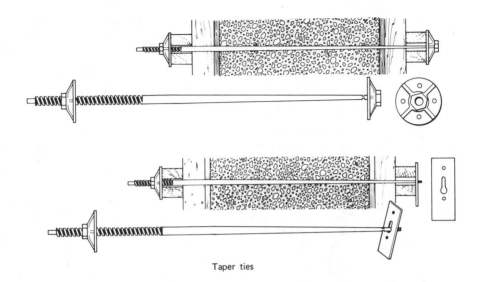

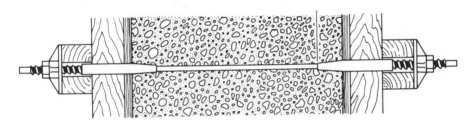

Taper ties

Waler rod or she—bolt

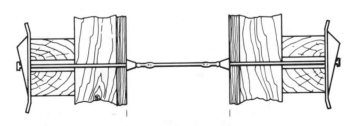

Snap tie

Figure 4.37 Form ties (The Burke Company).

ITEM 46   340 CY STRUCTURE CONC.

C 16.34 K$_{6+2}$ 1000 L 13.48 K$_{cr}$ 75 F 16.10		L	DAYS	M,R
S$_0$ GENERAL COSTS		⎰		⎱
CONC. 342.9 ×1.02 = 350 CY	30			10500
16.8 MFBM 2X	300			5040
17.7 ✓ HAVE	120			2120
5.2 MSF PLY	500			2600
8.0 ✓ HAVE	170			1360
218 CWT SHOR'G. 2 MOS.	436			870
173 ✓ WF 1 USE	1.50			260
8.2 C TIES 2 MOS.	40/C			655
1.0 C SHE-BOLTS 0.5 MO	✓			20
MINOR EXP. 0.343 MCY	3560			1220
MAKE 8003 SF	0.26	2080		—
ADD'L. FIN. 0.343 MCY	5180	1775		70
ADD'L. CURE 6 WKS.	90	540		—
SAND BLAST 271 SF	0.30	80		15
COMPRESSOR 2 MOS.	400			800
VIBRATORS ✓	200			400
DECK FIN. MACH. 4480 SF	0.05			225
GENERATOR 2 MOS.	250			500
DEVELOP WATER 100+150				250
HOPPERS				20
CURE RUGS				130
MISC. IRON				660
SWEEP 4.48 MSF	13	60		—
CLEAN UP 47.7 MFBM 43	12	2050		570
20 T 34	7	680		140
S$_{13}$ ABUT. + WW		—		
FG 216 SF	0.40	85	$\dfrac{2885}{1000×2/3}$	
SS 1436 SF H $^5/_{10}$	1.95	2800	=5 D	
42.8 POUR 44	5.25	230		
S$_{14}$ BENT F'ND.		—		
FG 200	0.28	55	$\dfrac{245}{1000×1/5}$	
3.5 CY ADD'L. CONC. 1	30	5	=1 D	105
SS 2 F'ND'S.	93	185		
16.7 POUR 17	4.45	75		

Figure 4.38 Estimate for T-girder bridge concrete (cont.).

ITEM 46 340 CY STR. CONC. (CONT'D.)					L	DAYS	M,R

					L	DAYS	M,R
	S_{15} BENT COL'S.				}	1030.	}
	SS 560 H2O			1.84	1030	1000×⅔	
17.8	POUR 18 CY		8.30	10.50	150	= 2 D	190
	S_{16} GIR'S. + SLAB				}		
	SS DECK B'LKH'D. 18 SF			5.30	95	17315	
	SS GIR ✓ 26 SF			1.89	50	1000	
	SS 6567 H4⅔			1.89	12410	= 18	
	SS 3411 SF SOF.			1.20	4090	+2+8	
	SS 158 SF DE H7"			4.25	670	=28D	
2656	POUR 271 CY		7.00	5.00	1900		1350
3429	S_{17} FALSEWORK				}		}
	SS 218 CWT 2B+E		8.90	0.60	1940	6150	130
	SS 20 FP			7	140	1000	—
	SS 20 16WF36 24'		40	29	800	=7 D	580
	SS 15.5 M.FBM		211	54	3270		835
					37245		31615
	INCL. PAYROLL EXP. (1.315)				48980		

$$\Sigma = \$80,595 \quad (\$235/CY)$$

TIME EST.	CONSEC.	CONC'T.
ON-OFF	2	2
ABUT'S.	5	3
BENT F'ND'S.	1	1
COL'S.	2	2
GIR.+SLAB	28	28
F'W'K.	7	5
REINF.	—	}
ABUT	1 }	
COLS	1 }	4
GIR+SLAB	6)	—
	53 D	45 D

Figure **4.39** Estimate for T-girder bridge concrete (cont.).

ITEM 46 WORK SHEET

		M,R
MISC. IRON		
44 PL 4 × 3/8 × 4 1 HOLE EA.	3	132
16 PL 2 1/2 × 3/8 × 12 ✓	4	64
80 LF 3/4" φ ROD	0.40	32
8 3/4" TURNBUCKLES	7	56
36 3/4" φ × 9" MB	1	36
8 3/4" φ × 2" ✓	0.50	4
40 1/2" φ × 46" TIE RODS	1.20	48
60 1/2" φ × 30" ✓	0.80	48
PORTABLE WELDER 8 HRS.	30	240
		660
MINOR EXPENSES		
NAILS 0.343 MCY @ 10.5 = 4 CWT	40	160
HAND TOOLS 0.343 MCY	500	170
ELECTRIC TOOLS 2 MOS.	70	140
FILING ✓	70	140
CURE. COMP (DECK) 30 GAL	1 + 40	70
FORM OIL 90 GAL	0.50 + 40	85
WEDGES 20 PR	0.40	8
CHAMFER 3440 LF	0.13	447
(0.343 MCY @ 3557)		1220
SAND BLAST		
ABUT'S. (2.5 × 32 + 4.75 × 2) 2	} 271 SF	
COL'S. 3 × 4 × 4		
POUR J'TS. 18 + 26		

Figure 4.40 Work sheet for T-girder bridge estimate.

ITEM 46. WORK SHEET

20 T FLAT BED TRUCK 30 K$_{cr}$ 75 FORK LIFT 30	L		M, R

ADD'L. FINISHING

   GRIND COL'S. + WW 0.784MSF 270   212
     14×18×2 + $\overline{5}$ ×14×4
   SACK OS GIR + DE   1.23 MSF 500   615
     140×3.83×2 +158
   FILL TIE HOLES

   7995/$_4$ + 40 (COL) = 2039 TOT.

    OS GIR 3.83×132 × ^{2}/$_4$ = 0.253 M 600   152
    OTHER 2039 − 253 = 1.786 M 300   536
  TOUCH UP 8.713 MSF   30   261
     (0.343 MCY @ 5178)   1776

CLEAN UP RENTAL

  TRUCKS 12 MFBM OR 20T STEEL PER LD.
    5 LDS. REQ'D. USE 2 TRUCKS
    30M/$_{40}$ MPH = 0.75 HR TRAVEL
            + 0.67 ✓ LOAD-UNLOAD
               1.42 ✓ CYCLE SAY 1½ HR
  HAUL COST PER LD. = 1.5×30   45
CRANE LOAD TIME 20 MIN
    30M/$_{30}$MPH = 1.00 HR TRAVEL
    LOAD TIME 4 @ 0.75 HR + 0.33 HR = 3.33 HRS
    3.33 + 1.00 = 4.33 HRS SAY 4.5 HR
    LOAD COST PER LOAD 4.5×75/$_5$ 68
FORK LIFT UNLOAD IN 20 MIN
    TRAVEL TIME ON-OFF ASSUME 0.67 HR
    UNLOAD TIME = LOAD TIME   3.33 HR
    UNLOAD COST PER LOAD 4×30/$_5$ 24
                                137

    137/$_{12}$ = $11.40 PER MFBM
    137/$_{20}$ = $6.85 PER TON

Figure 4.41 Work sheet for T-girder bridge estimate (continued).

Table 4.13. Unit Costs for T-Girder Bridges

Activity	Labor		Rental		Unit of Measure
	Unit Cost	Index	Unit Cost	Index	
Make girder form panels	$ 0.26	$K_{6+2} = 1000$	—		SF
SS girder forms and pour bulkheads	$ 1.89	$K_{6+2} = 1000$	—		SF
SS deck pour bulkheads[a]	$ 5.30	$K_{6+2} = 1000$	—		SF
SS net slab soffit	$ 1.20	$K_{6+2} = 1000$	—		SF
SS deck edge H6″–H7″	$ 4.25	$K_{6+2} = 1000$	—		SF
Pour girders and slab	$ 7.00	$K_{L+F} = 29.58$	$5.00	$K_{Cr} = 75$	CY
Pour girders only	$ 5.00	$K_L = 13.48$	$4.80	$K_{Cr} = 75$	CY
Pour deck slab only	$11.20	$K_{L+F} = 29.58$	$7.05	$K_{Cr} = 75$	CY

[a]Increase unit cost 30% when bulkhead is skewed with respect to the transverse reinforcing steel.

Note: The equipment costs for pouring assume normal access for the crane, and that the concrete can be deposited directly into the forms. When pours are large or average in size, the unit costs for the crane include 1 hour of travel time. On small pours, travel time will be additional.

with an assumed requirement of the specifications for a 10 calendar day curing period prior to commencing the stripping operations.

The number of working days required to complete the item is determined at the bottom of the estimate sheet, and amounts to 45 working days. Since 8 of these days represent a curing period, 2 months are allowed for rental periods in the general costs section.

4.11.2   Barrier Railing

We will assume the bridge has type 3-3 barrier railing on one side, and type 2 barrier railing on the other. The 3-3 designation indicates the concrete walkway is 3 feet wide. The pay length for each railing will be 168LF out-to-out of the concrete end posts. Details for concrete barrier rails were shown in Figure 4.29.

4.11.2.1   THE CONCRETE TAKE-OFF.   The take-off for the type 3 railing is shown in Figure 4.42. The S_{19} totals show the cubic yards of concrete in the walkway and the wall section separately, since two distinct operations requiring different unit costs are involved. In addition, the contact areas of the wall, walkway face, and end posts are listed.

As previously noted, when considerable amounts of barrier railing are required, 100 LF of railing wall is normally completed per day. In this case, we plan to make 82 LF of railing forms plus the forms for one end post, and to cut off the wall forms for re-use on the type 2 railing.

For this illustration, the areas of the required forms are converted in the general costs section into MSF of plyform and MFBM of 2X material. We note that we have this material, and will re-use material previously furnished for the bridge formwork.

The form tie holes to be filled consist of the holes in the end posts plus those on the back face of the wall. The holes on the front face will be closed at no additional cost during the float-finish operation.

An allowance is made in the minor expenses section for the use of driven studs at four foot centers to hold the forms to line.

The take-off for the type 2 barrier railing is shown in Figure 4.43, and the area of the forms to be modified is noted in the general costs section. No form ties pass through the type 2 wall, and only those holes in the end posts require patching.

4.11.2.2   THE BARRIER RAILING ESTIMATES.   The estimate sheets for the barrier railings are shown in Figures 4.44 and 4.45, and the unit costs for this work were given in Table 4.12.

TYPE 3 BARRIER RAILING	MISC,	POSTS	WALK	WALL	WALK CY	1. WALL CY
END POSTS $2(3\times1\times3.21/27)$						0.7
$2(3\times3.21+4.21\times1+0.33\times1.54)$		29				
WALL $A_S = 0.67(2.21+0.33)=1.70\,SF$						
$162\times1.70/27$ 162×4.75				770		10.2
WALK $2.92\times0.70\times168/27$					12.7	
$(168+6)0.7$			122			
$A_S = 0.75\times0.5=0.375\,SF$						
$0.375(28+5.5)/27$					0.5	
H EXC. 0.5 CY	1					
FG $28\times3 = 84\,SF$	1					
S_{19}	2	29	122	770	13.2	10.9
		MISC.	MAKE WALK	MAKE WALL	PLY	2X
GENERAL COSTS						
MAKE 82 LF 82×4.75				390		
$390+15 =405\,SF \times 1.17, 1.5$					0.5	0.6
MAKE 15 SF END POST		1				
MAKE $(82+3)0.7$ WALK			60			0.2
SAND BLAST $6\times1+162\times0.7$						
$+140\times3 = 540\,SF$		1				
SACK $162\times2.54+29 = 441\,SF$		1				
TIE HOLES $168/2 = 84, +1+8 = 93$		1			HAVE	
S_0	4	60	390		0.5	0.8
MINOR EXPENSES						
NAILS 0.024 X10.5 CWT 40						
HAND TOOLS 0.024 MCY 500	*68					
FORM OIL $92\frac{1}{150}=6$ GAL 0.50						
CHAMFER 3×82 LF 0.13						
DRIVE STUDS 35 EA. 0.30						

Figure 4.42 Take-off for Type 3 barrier railing.

TYPE 2 BARRIER RAILING	Σ SF	POSTS	LF	CY
END POSTS $2(3 \times 1 \times 2.25/27)$				0.5
$2(3 \times 2.25 + 3.25 \times 1 + 0.33 \times 0.58$	21	21		
WALL $A_s = 0.67(1.25 + 0.33) = 1.06$ SF				
$162 \times 1.06/27 \qquad 162 \times 2.89$	459		162	6.4
S$_{20}$	480	21	162	6.9
		MISC.		

GENERAL COSTS

MODIFY 390 SF		1		
SAND BLAST $6 \times 1 + 162 \times 0.7 = 120$ SF		1		
SACK $162 \times 1.58 + 21 = 277$ SF		1		
TIE HOLES 8		1		
S$_0$		4		

MINOR EXPENSES

NAILS 0.007 MCY × 10.5	40	
HAND TOOLS & INCIDENTAL 0.007 MCY	500	
FORM OIL 480/150 = 4 GAL	0.50	$35
CHAMFER $3 \times 82 \times 1/2$	0.13	
DRIVE STUDS 35 EA	0.30	

Figure 4.43 Take-off for Type 2 barrier railing.

ITEM 48 168 LF TYPE 3-3 BARRIER RAILING				
C 16.34 K_{6+2} =1000 L 13.48 K_{L+F} = 29.58 F 16.10		L	DAYS	M,R

			L	DAYS	M,R
	S_0 GENERAL COSTS				—
	24.1 x 1.02 = 24.75 CY	30			743
	MINOR EXP. 168 LF	0.40			67
	MAKE 390 SF WALL	1.17	456		
	✓ 60 SF WALK	0.25	15		
	✓ 15 SF END POST	1.17	18		
	SACK 0.441 MSF	500	220		10
	FILL 93 TIE HOLES	0.30	28		
	ADD'L. CURE 1 WK. END ½ x 50		25		
	SAND BLAST 540 SF	0.30	162		27
	SUBCONTRACT				
	168 LF METAL RAILING	10.50			1764
	24 CWT REINF. STEEL	25.10			602
	S_{19} BARRIER RAILING				
	H. EX. 0.5 CY	14	7	162	
	F G 84 SF	0.40	34	81	
	SS 770 SF WALL	1.60	1232	=2D	
	SS 122 SF WALK	2.50	305		
10.9	SS 29 SF POSTS	1.60	46		
13.2	POUR 11.25 CY WALL	43.50	490		
24.1	POUR 13.5 CY WALK	26	350		
			3388		3213
	INCL. PAYROLL EXP. (1.315)		4455		
	Σ = $7668 ($ 45.64/LF)				

Figure 4.44 Estimate for Type 3 barrier railing.

ITEM 47 168 LF TYPE 2 BARRIER RAILING			

$C\ 16.34$ $K_{6+2}=1000$ $L\ 13.48$ $K_{L+F}=29.58$ $F\ 16.10$		L	DAYS	M,R
S_0 GENERAL COSTS		$\}$		$\}$
6.9 × 1.02 = 7.25 CY	30			218
MINOR EXP. 168 LF	0.21	$\}$		35
MODIFY 390 SF	0.20	78		—
SACK 277 SF	0.50	138		6
FILL 8 TIE HOLES	0.30	2		—
ADD'L. CURE 1 WK. END	½×50	25		—
SAND BLAST 120 SF	0.30	36		6
SUBCONTRACT		$\}$		—
168 LF METAL RAILING	10.50			1764
15.6 CWT REINF. STEEL	25.10	$\}$		392
S_{20} BARRIER RAILING		$\}$		$\}$
SS 162 LF WALL	6.13	993	162	
SS 21 SF POSTS	1.60	34	81	
6.9 POUR 7.25 CY	43.50	315	=2 D	
		1621		2421
INCL. PAYROLL EXP. (1.315)		2132		
$\Sigma = \$4553$ ($\$27.10/_{LF}$)				

Figure 4.45 Estimate for Type 2 barrier railing.

Items listed in the general costs sections are taken from the general costs check list. In this case, no additional form lumber is required, and no additional charges are made for the compressor, vibrators, and curing blankets, or for developing power and water, because these costs were covered in the estimate for the bridge itself.

We have assumed that no expansion joint material is specified. When this material is required, the labor cost to set it is included in the unit cost to set and strip the barrier railing. In all cases, the costs to set and strip bulkheads at the ends of the walls and at construction or expansion joints are included in the unit costs to set and strip the barrier railing walls, and no additional charges are necessary for this work.

The cost to furnish and install the metal railing is a subcontract cost, and is normally included in the cost of the barrier railing. In this example we assume a cost of $10.50 per linear foot of the pay item. Since, due to the concrete end posts, the pay length for barrier railing is greater than the length of metal railing required, it is essential to know exactly what basis is being used by each subcontractor quoting metal railing.

In our estimates we assume the cost of the reinforcing steel to be included in the cost of the barrier railing. This is a subcontract cost, and may be quoted on a weight basis, on a linear foot basis, or as a lump sum.

To save time on bid day, the subcontract items may be processed on the estimate summary sheet, or assumed costs may be entered on the estimate sheets for the barrier railing and then corrected on the cut sheet as quotations are received.

The time estimates for the barrier railing are made by dividing the linear feet to be installed by our assumed rate of 81 linear feet per day. We will complete the walkway portion of the type 3 railing while constructing the type 2 railing. Under normal circumstances, it is not necessary to allow additional time for installing the metal rail, since this work is done concurrently with the final job cleanup or other items of work.

4.12 Box Girder Bridges

The box girder bridge is shown in Figure 4.46. Type 1 abutments with the dimensions shown in Figure 4.27 are used. It is assumed that the job conditions and location are the same as those described in connection with the estimate for the retaining walls.

4.12.1 Structure Concrete

It is also assumed that the box girder bridge constitutes the total structure concrete item for the project.

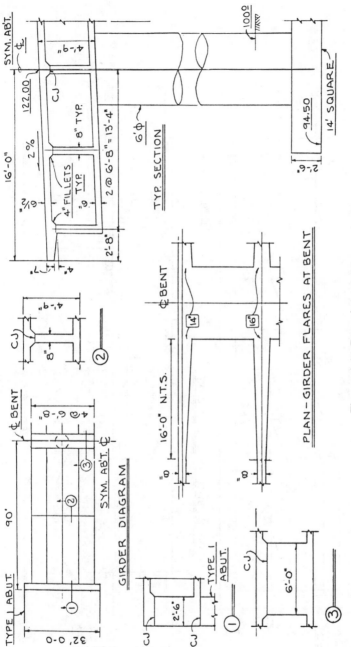

Figure 4.46 Box girder bridge.

163

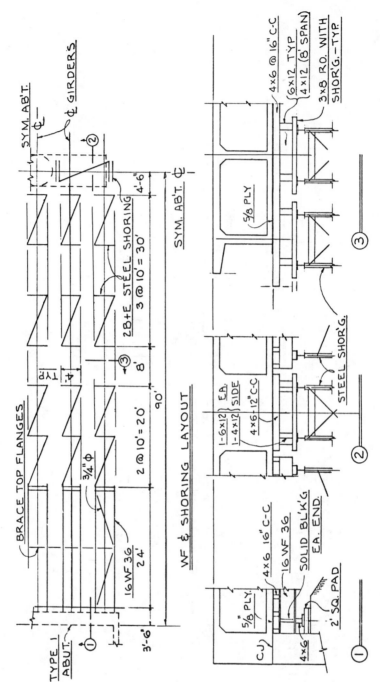

Figure 4.47 Falsework for box girder bridge.

4.12.1.1 THE CONCRETE TAKE-OFF. The concrete take-off is shown in Figures 4.48 to 4.50, and includes the take-off of the required falsework shown in Figure 4.47. The falsework is based on a preliminary design, and in practice the falsework take-off would be made from sketches on the design sheets. The preliminary falsework design requires very little time. The toe of the abutments slopes is located, and the shoring towers are spaced so that the footing pads for the end frames nearest to the abutments are, preferably, on the flat surface.

When circumstances require these pads to be located on the toe of the abutment slopes, the toe is removed and later replaced. It is possible to raise the elevation of these end pads by replacing the bottom base frames with inverted extension frames. This procedure permits the end pads to be located on the abutment slopes but requires hand excavation to set the pads; it increases the cost to set and strip the shoring, and should be avoided when possible. It is seldom practical to set an entire shoring tower on an abutment slope.

Since the abutments for this bridge are identical to those previously used for the T-girder bridge, the S_{13} total from Figure 4.33 is transferred directly to this take-off.

To save time and to reduce the likelihood of error, the calculations for the volume of concrete involved in S_{23} and S_{24} do not conform exactly with the true location of the construction joints. The abutment diaphragms are assumed to extend to the top of the top slab, and the height of the girders, bent diaphragm, and intermediate diaphragms are taken as the clear distance between the top and bottom slabs.

The calculations for the contact areas of the required forms are based on the true locations of the construction joints. No deductions are made in the areas of the girder forms at their junction with the intermediate diaphragms, and the effect of the girder flares is neglected in determining the contact area of the bent diaphragm forms. The additional thickness of the deck overhangs is also neglected in determining the contact areas of the abutment diaphragm forms.

In our case the method used to calculate the concrete volumes increases the true S_{23} volume by 14 CY and decreases the true S_{24} volume by the same amount. Because of differences in the pour costs, an error of $86 is introduced into the estimate, but this discrepancy is on the low side and is considered insignificant.

The areas of the required slab and girder pour bulkheads are noted in the miscellaneous column of section S_{23}, since we plan to make two separate pours for this work.

In section S_{25}, the 4 × 6 is listed separately from the 4 × 12 and 6 × 12 stringers, since the cost to buy these materials would be different. Upon

BOX GIRDER BRIDGE		FG	H$^5/_{10}$	L. CY
ABUTMENTS & WINGWALLS				
S_{13} (FIG. 4-33)		216	1436	42.8
		MISC	FG	CY
BENT F'ND.				
$14 \times 14 \times {}^{2.5}/_{27}$			196	18.2
POUR NEAT ADD'L. CONC. 3.5 CY		1		
$15.3 \times 15.3 \times {}^{2.5}/_{27} - 18.2$				
S_{21}		1	196	18.2
			6'ϕ H 20$\frac{2}{8}$	CY
COLUMN H=122.00−4.75−97.00=20.25				
$6^2 \times {}^{\pi}\!/_4 \times 20.25/_{27}$ S_{22}			1	21.2
	MISC.	H3$\frac{4}{}$	SOF	CY
BOT. SLAB + GIR'S.				
SLAB				
$27.33 \times 175 \times {}^{0.5}/_{27}$				88.6
$27.33 \times 175 - {}^{\pi}\!/_4 \times 36$			4755	
GIR'S. L$_G$=175−6 = 169'				
h$_c$ = 4.75−1.04 =3.71' h$_f$ = 3.71−0.33 = 3.38' INT.				
$5(169 \times 0.667 \times {}^{3.71}/_{27})$				77.4
$2(3.83 \times 169) + 8(3.38 \times 169)$		5864		
FLARES ADD'L.				
$4(3.71 \times 0.\overline{25} \times {}^{16}/_{27}) + 6(3.71 \times 0.\overline{33} \times {}^{16}/_{27})$				6.6
ABUT. DIA.				
$2.5 \times 4.75 \times {}^{32}/_{27}$				14.1
$4.75 \times 30 + 4.21 \times 32 + 4.75 \times 5$				
$-27.33 \times 0.5 - 0.67 \times 3.71 \times 5$		275		
1 ADD'L. ABUT. DIA.		275		14.1
BENT DIA.				
$6 \times 3.71 \times {}^{27.33}/_{27}$				22.5
$2(27.33 \times 3.71) + 2(6 \times 3.83) - 10(6.67 \times 3.71)$		224		

Figure 4.48 Concrete take-off for box girder bridge.

BOX GIRDER BRIDGE (CONT'D.)

	MISC	H 3½	SOF	CY
INTERMEDIATE DIA'S.				
8 (6.00 × 3.71 × 0.667/27)				4.4
8 (6.00 × 3.71 × 2)		356		
POUR BULKHEADS				
SLAB 27.33 × 0.5 = 14 SF ⎫				
GIR'S. 5 × 0.67 × 3.38 = 11 SF ⎭	1			
S_{23}	1	6994	4755	227.7

		H7" DE	OH SOF	LOST SOF	CY
TOP SLAB					
32 × 175 × 0.542/27					112.4
169 × 6.00 × 4				4056	
ADD'L. @ OH (2.33 × 0.21 × 175/27) 2					6.3
175 × 2.33 × 2			816		
175 × 0.583 × 2		204			
FILLETS L = 8×169 + 8×4×6 = 1544'					
1544 × 0.333² × ½ × 1/27					3.2
1544 (0.333√2 − 0.333)				213	
S_{24}		204	816	4269	121.9

$$\sum_{21}^{24} S + S_{13} = 431.8 \text{ CY}$$

	MISC.	2B+E	SS	4×6	4×12 6×12
FALSEWORK					
BENT IN 2B+E 2 @ 12.32 CWT		24.6			
4×12 4/12			0.19		0.19
6×12 4/12			0.29		0.29
4×6 28/6			0.34	0.34	
GIR'S.					
IN 2B+E 10 @ 13.28 CWT		132.8			
N 2B+E 20 @ 6.58 CWT		131.6			
END CAPS 4×6 68LF			0.14	0.14	
FLAT PADS (4 SF) 20 EA	1		—	0.3	
6×12 100/12			7.20		7.20
4×12 20/10			0.80		0.80
16 WF 36 20/26 = 187.7 CWT	1				
TIES 4×6 60 LF			—	0.12	
RODS ADD'L. (HAVE)	1				

Figure 4.49 Concrete take-off for box girder bridge.

BOX GIRDER BRIDGE (CONT'D)				3. 4×12
	MISC 2B+E	SS	4×6	6×12

```
 4×6   90-25-3 = 84.5, ÷1.33 = 63.5
        (65+1) 2 = 132 LINES
     ┌─ 14'⌐ ─┬──── 10'⌐ ───┬─ 14'⌐ ─┐
   | 4' |2.7| 4' |2.7| 4' |2.7| 4' |2.7| 4' |
 4×6  264/14   132/10
```

			SS	4×6	6×12
4×6			10.03	10.03	
S₂₅	3	289.0	19.0	10.9	8.5
				HAVE	HAVE
			LOST		
	MISC	MAKE	HAVE	PLY	PLY
					2X

GENERAL COSTS

	MISC	MAKE	HAVE	PLY	2X
S₁₃ ABUT'S. MAKE 1 718 × 1.17, 1.75		718		0.84	1.26
S₂₂ COL. 1 EA 6'φ × 21.25'	1				
S₂₃ BOT. SL. + GIR'S. SOF. 4755 × 1.17				5.56	
OS GIR COMPLETE					
2 × 3.83 × 169 = 1295					
110 LF I.S. GIR					
(3.58 × 110) 8 = 2974					
2 ABUT DIA 550					
BENT DIA 224					
½ INT. DIA 178					
✱ 1.17, 1.75 × 5221		5221		6.11	9.14
S₂₄ DECK SLAB					
DE 204 × 1.17, 1.75		204		0.24	0.36
OH SOF 816 × 1.17, 4.8				0.95	3.9
LOST SOF 4269 × 1.17, 1.35			5.0		5.8
RE-USE S₁₃ (0.84, 1.26) × 2/3				(0.56)	(0.84)
✓ S₂₃					
2974 + ½ × 550 + 224 + 178 = 3651					
3651 × 2/3 × 1.75					(4.3)
✓ × 1.17 × 10% ±			(0.5)		
S₂₅ FALSEW'K. I HAVE			19.4		
289 CWT 2B+E, 187 CWT 16 WF	2				
✱ TAPER TIES 5221 ÷ 8 = 653	1				
S₀	4	6143	19.4	4.5	15.3

(HAVE 6.0) in 4×6 column → 13.1; (HAVE) in SS column → 4.5

Figure 4.50 Concrete take-off for box girder bridge.

Figure 4H Box Girder Bridge. Bakersfield, California. Contractor: Tumblin Company.

determining the amounts required, however, we note that we have sufficient material in our yard. In taking off the 4 × 6, we add one additional line of 4 × 6's for each span, and these 4 × 6's will be used under the intermediate diaphragms.

The flat pads required at the end caps are noted and the requisite material is listed. We assume we have these pads in our yard, and no charge is necessary to make them.

As noted in Table 4.5, and Figure 4.16, no additional charges are made for setting, stripping, or hauling the footing pads required for steel shoring or for wooden bent cap falsework. In both cases the pads are set on essentially flat surfaces, no "slope pads" with their attendant hand excavation are involved, and the costs to set and strip the pads are included in the unit costs to set and strip the falsework.

When wooden falsework is used in connection with slab or girder bridges, the cost to set and strip the pads is based on separate unit costs since such pads may be set on flat or sloped surfaces. Wooden falsework may be used on abutment slopes when steel beams are not available, but the cost will be greater than that realized when such falsework is eliminated through the use of steel beams.

In our case, we assume we have the necessary pads with the steel shoring and no additional charges for furnishing them will be required.

In section S_0 the areas from sections S_{13} and S_{21} to S_{24} are converted into MFBM of 2 × 4 and MSF of plywood, and the areas of the form panels to be made are listed in the "make" column.

We plan to make form panels for all the exterior girder faces, since these forms remain in place until the top slab is stripped. Since we plan to make two pours, we will make forms for only 110LF of each interior girder face.

An allowance of 4.8 MFBM of lumber per MSF of deck overhang soffit is used, based on the determination made in connection with the T-girder bridge. The allowance of 1.35 MFBM per MSF of lost soffit is based on many past designs.

Allowances are made for re-use of form materials from sections S_{13} and S_{23}. In constructing box girder bridges, the interior girder and diaphragm forms are stripped, but the interior soffit forms remain in place and are lost.

In this instance, we estimate that 4.3 MFBM of material will be available for use in the lost soffit formwork. Although we have more than 4MSF of plyform available, we consider this too valuable to lose, since we have scrap plyform in our yard. Therefore, we allow only about ten percent of this material for re-use and plan to use the scrap plyform for the remainder.

If scrap plyform were not available, a decision would be required regarding re-use of the ⅝ inch material from the girders versus the purchase of the cheapest grade of ⅜ inch interior plywood. In many cases ⅜ inch plyscord is used in lost soffit formwork.

4.12.1.2   THE STRUCTURE CONCRETE ESTIMATE.   The structure concrete estimate is shown in Figures 4.51 and 4.52, and the unit costs for this work are given in Tables 4.5, 4.6, 4.10, 4.11, and 4.14.

The allowance for additional curing depends on the arrangements we can make to have a workman tend to the cure over the weekends. In our case we have assumed that a nearby water supply is available, and have allowed $100 to purchase the water and $150 for water hoses and nozzles. On many jobs a rental charge to cover the cost of water trucks, storage tanks, and pumps may be required.

The small allowance for hoppers is made to cover the collection hopper and elephant trunks required to pour the bent column, and we assume we have this equipment in our yard.

The miscellaneous iron item allows 4 hours of welder time to modify the rod bracing for the top flanges of the wide flange beams. We assume we have rods used on past work in our yard.

The 41.5 MFBM of material noted in the cleanup item represents the sum of the MFBM and MSF of material listed to be furnished, less the 10.8 MFBM of material in the lost soffit formwork. For this calculation one MSF is assumed to equal 1 MFBM.

ITEM 37 430 CY STRUCTURE CONC.

C 16.34 K6+2 1000 L 13.48 F 16.10 Kcr 75		L	DAYS	M,R
S₀ GENERAL COSTS				}
CONC. 431.8×1.02=441 CY	30			13230
15.3 MFBM 2X	300			4590
19.4 ˝ HAVE	120			2330
7.1 MSF PLY	500			3550
6.0 ˝ ˝ HAVE	170			1020
4.5 ˝ LOST PLY HAVE	50			225
289 CWT 2B+E 2 MOS.	578			1160
187 CWT WF I USE.	1.50			280
COL. FORM 6'∅×24' I WK	450			450
6.5 C TIES 2 MOS.	40/c			520
MINOR EXP. 0.432 MCY	2370			1030
MAKE 6143 SF	0.26	1600		—
ADD'L. FIN. 0.432 MCY	3630	1570		60
ADD'L. CURE 5 WKS	90	450		—
SAND BLAST 0.432 MCY	872	375		65
COMPRESSOR 2 MOS	400			800
VIBRATORS ˝ ˝	200			400
DECK FIN. MACH. 5760 SF	0.05			290
GENERATOR 2 MOS.	200			400
DEVELOP WATER 100+150				250
HOPPERS				20
CURE RUGS				130
MISC. IRON 4 HR WELDER	30			120
BOX GIR. DOBIES 0.228 MCY	700	160		10
SWEEP 5.76 MSF	13	75		
CLEAN UP 41.5 MFBM 43	12	1780		500
23.8 T 34	7	810		165
S₁₃ ABUT. + WW			2885	
FG 216 SF	0.40	85	1000×⅔	
SS 1436 SF H ⁵/₁₀	1.95	2800	=5 D	
POUR 44 CY	5.25	230		
S₂₁ BENT F'ND.			160	
FG 196 SF	0.28	55	1000×⅓	
3.5 CY ADD'L. CONC. I	30	5	=1 D	105
SS I F'ND.	110%×93	100		
POUR 19 CY	4.45	85		

42.8

Figure 4.51 Estimate for box girder bridge.

ITEM 37 430 CY STR. CONC. (CONT'D)

			L	DAYS	M,R
	S₂₂ BENT COL.		⎰		⎰
	SS I COL. 360 150		360	$\dfrac{360}{1000 \times 2/3}$	150
	CRANE ON-OFF ADD'L 2 HR 75		—		150
21.2	POUR 22 8.30 10.50		185	=1D	230
	S₂₃ BOT. SLAB+GIR'S.		⎰		⎰
	SS SLAB BULKH'D. 14 SF 5.30		75	$\dfrac{8130}{1000}$	
	SS GIR BULKH'DS. 11 SF 1.15		15		
	SS 6994 SF H3½ 1.15		8040	=9,+1	⎰
227.7	POUR 232 CY 7.75 5.80		1800	10D	1350
	S₂₄ TOP SLAB		⎰	$\dfrac{6615}{1000}$	
	SS 204 SF DE 4.25		865		
	SS 5085 SF (LOST+OH) SOF. 1.13		5750	=7,+1+8	
121.9 431.8′	POUR 124 CY 11.20 7.05		1390	16D	875
	S₂₅ FALSEWORK		⎰		⎰
	SS 289 CWT 2B+E 8.90 0.60		2570	$\dfrac{7520}{1000}$	175
	SS 20 16WF 36-26′ 40 29		800		580
	SS 20 F PADS 7		140	=8D	—
	SS 19.0 MFBM 211 54		4010		1030
			36180		36240
	INCL. PAYROLL EXP. (1.315)		47575		

$\Sigma = \$\,83,815$ ($\$194/CY$)

TIME EST.	CONSEC.	CONC'T.
ON-OFF	3	3
ABUT'S.	5	3
BENT F'ND + COL	2	2
BOT. SL. + GIR'S.	10	10
TOP SLAB	16	16
FALSEW'K	8	5
REINF.		
ABUT'S	1 ⎱	
COL	1 ⎰	6
BOT SL. + GIR'S	5	
TOP SL.	3 ⎱	
	54 D	45 D

Figure 4.52 Estimate for box girder bridge.

Table 4.14. Unit Costs for Box Girder Bridges

Activity	Labor		Rental		Unit of Measure
	Unit Cost	Index	Unit Cost	Index	
Additional finishing	$3630	$K_{L+F} = 29.58$	—		MCY
Sandblast	$ 872	$K_L = 13.48$	—		MCY
Additional cure	$ 775	$K_L = 13.48$	[a]		MCY
Make girder forms	$2770	$K_{6+2} = 1000$	—		MCY
Or, make girder forms	$ 0.26	$K_{6+2} = 1000$	—		SF
SS girder forms and pour bulkheads	$ 1.15	$K_{6+2} = 1000$	$0.06	$K_{Cr} = 75$	SF
SS slab pour bulkheads[b]	$ 5.30	$K_{6+2} = 1000$	—		SF
Pour bottom slab plus girders	$ 7.75	$K_{L+F} = 29.58$	$5.80	$K_{Cr} = 75$	CY
SS lost soffit and OH soffit[c]	$ 1.13	$K_{6+2} = 1000$	—		SF
SS deck edge H6"–H7"	$ 4.25	$K_{6+2} = 1000$	—		SF
Pour deck slab	$11.20	$K_{L+F} = 29.58$	$7.05	$K_{Cr} = 75$	CY

[a]Rental cost will depend on job conditions. See text.
[b]Increase unit cost 30 percent when bulkhead is skewed with respect to the transverse reinforcing steel.
[c]Increase unit cost 25 percent on skewed bridges when skew is greater than 20 degrees.

Note: The equipment costs for pouring assume normal access for the crane, and that the concrete can be deposited directly into the forms. When pours are large or average in size, the unit costs for the crane include 1 hour of travel time. On small pours, travel time will be additional.

173

Figure 4I Box Girder Bridges, Kern County, California. Contractor: Tumblin Company.

In section S_{21} we increase the unit cost to set and strip the bent foundation by 10 percent because our foundation is somewhat larger than average.

The unit costs shown for setting and stripping the girder forms include the costs to set and strip formed drains through the bottom slab. One or two 4-inch-square openings are normally specified for each box girder cell.

When utility openings through the diaphragms are specified, the unit cost to set and strip the necessary forms will be about 5 times the unit cost used to set and strip the girder forms.

When hinges are specified, the unit cost for setting and stripping the hinges will be the same as the unit cost for setting and stripping the girders. An additional allowance must be made to furnish and place neoprene pads, hinge armor, or hinge tie rods when required.

In section S_{23}, no charge is made for setting and stripping the plyform under the bottom slab. This material is located directly on the falsework and the cost to set and strip it is included, as always, in the unit costs to set and strip the falsework.

The time required to complete the various operations is determined in the "days" column. A two-carpenter crew is used on the bent foundations, four-carpenter crews are used on the abutment and column work, and six-carpenter crews are used for the bottom slab plus girders, top slab, and falsework.

One additional working day is added to the bottom slab plus girder time to allow for the first pour. Nine working days are added to the time required to set and strip the top slab; one working day for the slab pour, and eight working days to comply with a specified ten-calendar-day delay before starting the stripping operation.

The estimated time to complete the item is 45 working days, as shown in Figure 4.52.

4.13 Prestressed Box Girder Bridges

The procedures and unit costs necessary to estimate the cost of prestressed box girder bridges are identical to those described above except that some additional factors must be considered.

Although the specifications must be checked in each case, there is normally a single, separate job item for prestressing cast-in-place concrete, and this item will cover all of the prestressing required on the project.

Prestressing is done by subcontractors, and different subcontractors often quote different prestressing systems.

The subcontract price usually includes the cost to furnish design calculations relating to the prestressing system being used, as well as the complete cost to furnish and install the necessary ducts, tendons, and vents, including the costs to tension and grout the tendons.

4.13.1 Work Area and Water Supply

Prestressing subcontractors require sufficient work areas to enable them to stretch out and measure the ducts and tendons. On a typical bridge job this does not entail any additional cost to the general contractor, since the bridge approach roadways are available for this purpose.

In addition, the general contractor may be required to furnish the water required for the grouting operation.

4.13.2 Additional Reinforcing Steel

The specifications may require additional longitudinal reinforcing steel to be draped in the girder stems to facilitate the securing of the ducts. The cost of such reinforcing steel is frequently included in the prestressing item, but is usually excluded by the prestressing subcontractor. In such cases the cost to furnish and place this reinforcing represents an additional cost to the general contractor.

4.13.3   Vents

Vent pipes may be required. The vents are made of ½-inch-diameter iron or plastic pipe, and extend from the ducts upward through the top slab. After the tendons have been grouted, the vent pipes are cut off 1 inch below the top surface of the slab, and the deck is patched.

The general contractor normally blocks out the top slab to a depth of 1 inch prior to the pour, and subsequently cuts the vent pipes and patches the blocked out portion of the deck. Based on $K_{6+2} = 1000$, the cost for this work will run about $60 per girder per set of vents.

4.13.4   Girder Bulkheads

On prestressed box girder bridges, the ends of the girders normally extend a minimum of 6 inches beyond the outer face of the abutment diaphragms. On skewed bridges, the girders are extended as required to enable the end bulkhead to be set perpendicular to the girder center line. In addition, a section through the end bulkhead itself will be stepped or parabolic in shape. This is necessary to obtain bearing surfaces for the tendon locking devices, which will be perpendicular to the draped tendons. The details for the end bulkheads vary with the prestressing system being used, and the exact dimensions are furnished by the prestressing subcontractor.

The unit cost to make, set, and strip the formwork for the extended girder sides and ends runs about six times the unit cost used to set and strip the typical girder forms, and this cost will include the cost of laying out and drilling the holes required in the end bulkheads for the ducts.

After the bridge has been prestressed, the plans may require the locked tendon ends to be encased in concrete. The unit costs shown in Table 4.11 for curtain walls may be used to arrive at an estimated cost for this work.

4.13.5   Falsework During the Tensioning Operation

The box girder bridge will decrease in length and will rise, or hog up, at midspans as the girders are tensioned. The maximum horizontal movement from the reduction in length will occur at each end of the section being tensioned, or, typically, at the abutments.

As the bridge becomes shorter, the X braces in those shoring towers nearest the abutments will start to bow, and must be removed as necessary. Since the load on the towers is being reduced as the girders are being tensioned, there is no danger involved, but the operation should be supervised by a competent employee. The cost of the workman used to remove the X braces represents an additional cost to the general contractor.

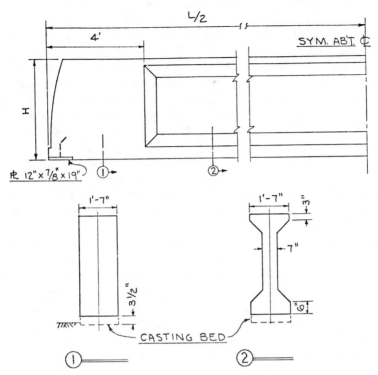

Figure 4.53 Precast concrete girders.

4.14 Precast Concrete Girders

A typical precast concrete girder is shown in Figure 4.53. Such girders are frequently used in constructing highway bridges. Their heights normally vary from 2.75 to 5.5 feet, and their lengths range from 50 to 120 feet.

When precast concrete girders are required, separate job items are set up for furnishing and erecting them. Thus, there may be one item for furnishing 15 girders 55 to 60 feet in length, a second item for furnishing 12 girders 60 to 65 feet in length, and a third item for erecting 27 girders 55 to 65 feet in length.

Both the furnishing and the erecting items are bid on a price per girder basis, and the price for furnishing the girders normally includes the costs for prestressing them.

Except for projects in remote locations, it is usually cheaper to have precast concrete girders furnished by precast concrete manufacturers. Savings are realized through the use of long casting beds which are set up for pretensioning, use of metal forms, availability of overhead gantry cranes, and use of sophisticated concrete mixing, placing, and vibrating equipment. In addition, the girders are usually steam cured.

Manufactured girders should always be erected by the manufacturer, since it is never wise to split the responsibility for a product unnecessarily.

Manufactured girders up to about 85 feet long are delivered to the job in one piece by trucks. Delivery is scheduled to fit the erection sequence, and the girders are taken from the trucks and placed in position in one operation.

Precast concrete I-girders over 85 feet long require special handling during the trucking operation to prevent buckling. Because of the added costs and risks associated with hauling the longer girders, manufacturers often fabricate them in two pieces, which are joined and post tensioned at the erection site.

The erection is considered to be completed when the girder has been set in its plan location and the slings have been removed. It is the general contractor's responsibility to tie off and brace the positioned girders prior to the removal of the slings. Unit costs for this work are shown in Table 4.15.

Care and good judgment must be exercised during the girder bracing operation, since, until the girder diaphragms are poured, the girders are somewhat unstable and are vulnerable to earthquake and, in the case of longer girders, to buckling.

The manufacturer's quoted price per girder should include the cost of furnishing holes through the exterior girder webs as required for attaching the deck overhang brackets, as well as the cost to furnish the inserts and holes necessary to tie the poured-in-place diaphragms to the girders. (See Figure 4.54). In addition, the manufacturer's price should include the cost of sandblasting the tops of the girders.

In all cases, the general contractor must contact each girder manufacturer before the bid date to determine exactly what will be included in each of their quotations.

4.14.1 Job Fabricated Precast Girders

Precast girders, when fabricated on the job, are normally post tensioned. In such cases the prestressing work is usually performed by a prestressing subcontractor whose price will include the costs to furnish design calculations relating to the prestressing system being used, to furnish and install the necessary ducts and tendons, and to tension and grout the tendons.

The deformed-bar reinforcing required for job-fabricated girders is fur-

Table 4.15. Unit Costs for Precast Prestressed Concrete I-Girders

Activity	Labor Unit Cost	Labor Index	Rental Unit Cost	Rental Index	Unit of Measure
FG, SS, P & F casting bed slabs	$ 1.34	K_{L+F} = 29.58	—		SF
Make side forms H < 4'	$ 4.00	K_{6+2} = 1000	—		LF[a]
H4'–H5.5'	$ 1.00	K_{6+2} = 1000	—		SF
Make end bulkheads	$ 5.65	K_{6+2} = 1000	—		SF
SS side forms H2.75'	$ 0.70	K_{6+2} = 1000	—		SF
H4'–H5.5'	$ 1.05	K_{6+2} = 1000	—		SF
SS end bulkheads	$ 4.75	K_{6+2} = 1000	—		SF
Form holes or set inserts	$ 2.90	K_{6+2} = 1000	—		EA
Set b'ng plates or cable loops	$ 1.50	K_{6+2} = 1000	—		EA
Pour girders	$10.50	K_{L+F} = 29.58	—		CY
Erect girders L50'–L85'[b]					
Crane rental	—		$1.58	K_{Cr} = 75	LF
Truck rental	—		$0.63	K_{Tr} = 30	LF
Rigger labor	$ 0.72	K_R = 17.00	—		LF
Tie off girders	$ 0.66	K_{6+2} = 1000	—		LF

[a]Per LF of form panels required for each side.

[b]See text for assumed conditions. On-and-off charges for the equipment and the cost to rig cranes, when required, are additional. If girders are cast at the erection site, eliminate the truck cost and reduce crane rigger costs by 50 percent.

Note: Unit costs to set neoprene pads and celotex are shown on Table 4.16.

179

nished and placed by the subcontractor used on the bridge work, but the cost is normally included in the cost of furnishing the girders.

Although the specifications must be checked in each case, the cost to furnish embedded miscellaneous iron such as bearing plates is usually included in the costs to furnish the girders.

4.14.1.1 THE CONCRETE TAKE-OFF.   The unit costs to fabricate precast girders at the job site are shown in Table 4.15, and the determination of the quantities relating to the activities listed constitutes the major portion of the concrete take-off. The following items must also be considered in estimating the cost of precast girders.

4.14.1.2 THE CASTING YARD.   The casting yard site must be obtained and it should be located as close to the point of erection as possible. The number of casting beds required depends on the girder erection schedule, and in many cases some or all of the casting beds may be used more than once. Two lines of casting beds may be set 7 feet center-to-center, with 20-foot-wide roadways at each side, and this pattern repeated as necessary.

The casting beds should be set low enough to permit the girders to be poured directly from the transit-mix trucks.

The costs to grade the site as well as to dispose of the concrete casting beds on completion must be included in the cost to furnish the girders. The costs for such work are discussed in the chapters on earthwork.

The lengths of the girders will decrease during the tensioning operation, the mid-portions of the girders will rise up off the casting beds, and the girders will be bearing only at their extreme ends. A short piece of plywood is embedded in the top surface of the casting bed at the girder ends to prevent damage. The hogged-up girders may require temporary ties to ensure stability. The cost of this work is nominal, and is included in the costs of the other activities.

4.14.1.3 GIRDER ERECTION.   The unit costs shown for erecting girders are based on using two cranes for each girder. It is assumed that the bridge is low enough to permit the cranes to erect the girders while setting below the bridge deck, and that normal access is available to this area for the trucks and cranes.

It is assumed further that the casting yard is approximately 400 feet from the erection site, and that the job conditions are such that the cranes cannot walk the girders to the point of erection. Two trucks are used, and the girders are loaded onto the trucks and subsequently erected with the same cranes.

In all crane work, it is imperative that the cranes selected have ample lifting capacities to handle the existing loads when their booms are lowered, as required by the job conditions. As a rule of thumb, the cranes selected should furnish twice the lifting capacity theoretically required.

On bridges having large skews, it may be impossible to erect girders with only two cranes. In such cases it may be necessary to transfer one girder end to a third crane located behind the bridge abutment.

When erection conditions are difficult, a plan view of the proposed crane is drawn on tracing paper, to the proper scale, and this model is shifted over the plan view of the girders and the abutment fill to determine the required horizontal projection of the boom.

The hauling and the erection of I-girders over 100 feet long present special problems. Even though great care has been exercised in arranging the ducts, such girders may bow into flat C or S shapes, when tensioned. In such cases lateral hog rod bracing may be required near the bottom flange before moving the girders from the casting beds, and this bracing must remain in place until the girders have been set and tied off. The costs to furnish this bracing, and the additional time required to place and remove it, increase the costs to furnish and erect long girders.

Due to the difficulties involved, the costs of cranes, trucks, and riggers required to erect 120 foot girders may run from 3 to 3½ times the costs shown in Table 4.15 for the shorter girders, and the cost to tie off and brace the 120-foot girders will run about twice those shown.

Flagmen and/or detours may be required during the erection operation. Flagmen always constitute an added expense, but the detour work may be paid for at the bid unit prices for the work involved.

4.14.1.4 GENERAL COSTS. The general costs section includes those items from the check list that have not been covered in the estimate for the structure concrete item.

The amount of forms to be made depends on the job schedule and on the number of girders required. 2½ sets of forms will enable two girders 60 to 80 feet long to be completed per day, or one girder 120 feet long to be completed on alternate days.

When forms are reused 12 or more times, it may be necessary to grind the girder sides lightly for uniform appearance. In all cases the exterior face of the exterior girders is sacked, together with the deck edge after the bridge deck has been completed.

The girders are easily cured with soaker hoses and polyethylene film.

The top surfaces of the girders must be sandblasted, and when large areas are involved, this work is done by sandblasting subcontractors.

Special vibrators are required for concrete girders because of the reduced clearances at the ducts. Two vibrators with ¾-inch-diameter spuds will be required for use in the worst places. Heavy vibration is necessary since high strengths and low slumps are involved.

Care must be taken to ensure that the forms do not rise during the pour. When such rise does occur, the bottom chamfers will require additional finishing, which may cost as much as $0.85 per linear foot of girder, based on $K_F = 16.10$.

A dependable bond breaker is required on the top surface of the casting bed. During rainy weather, it may be necessary to lay plywood strips on the casting bed to prevent adhesion.

A loop of wire rope is embedded at each end of the girder for lifting. These loops are located on the top surface, and are on the girder center line. The cost to furnish the wire rope used is included in the general costs section.

4.15   Precast Concrete Girder Bridge Decks

A precast concrete girder bridge is shown in Figure 4.54, and the deck soffit forms are shown in Figure 4.55. The concrete take-off for the bridge deck is shown in Figures 4.56 and 4.57, and includes the take-off for the soffit forms. Unit costs for the bridge deck work are given in Table 4.16.

The following discussion is limited to the costs of constructing precast concrete girder bridge decks, since take-offs and estimates for abutments, bents, and complete bridges have been illustrated in previous sections. The furnishing and erecting of the girders themselves is a separate job item, as was discussed in the preceding section.

4.15.1   Unit Costs

The unit cost for pouring the diaphragms is based on doing this work with concrete buggies running on the deck soffit forms.

The cost to make up the deck overhang brackets is carried separately, since these brackets may be re-used when several similar bridges are being built.

The unit cost to set and strip the soffit forms is applied to the gross area of the deck slab and no deductions are made for the plan areas of the girders or diaphragms. The unit cost to set and strip the soffit forms includes the costs to set and strip all the formwork shown in Figure 4.55, including the deck overhang brackets, hangers, rods, plywood, and chamfer strips.

Figure 4J Precast I-Girder Bridges, Beardsley Canal. Contractor: Tumblin Company.

Figure 4K Precast I-Girder Bridges, U.S. Highway 99. Contractor: Tumblin Company.

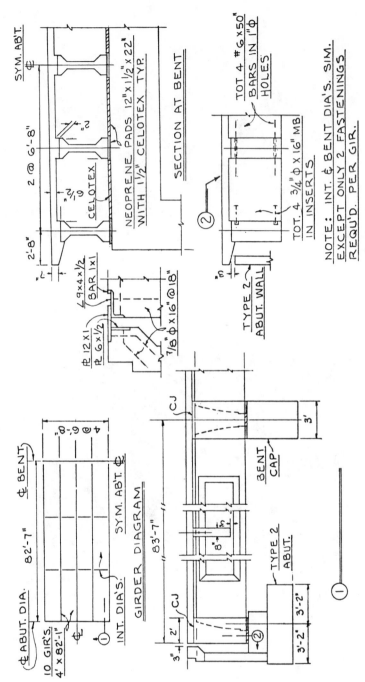

Figure 4.54 Precast concrete girder bridge.

184

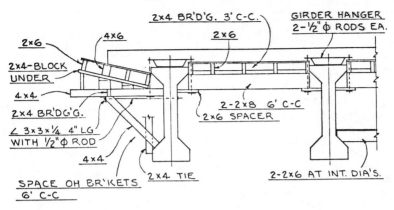

Figure 4.55 Soffit forms for precast concrete girder bridge.

4.15.2   The Concrete Take-off

The take-off for the general costs section is not illustrated, since it depends on the total amount of structure concrete involved.

Figure 4.54 shows the deck soffit to be 2 inches above the top of the girders. This dimension varies, in practice, and an approximate average value must be calculated. The calculation is based on the given elevations for the girder bearings at the abutments and at the bent cap, the given centerline profile, and the given camber diagrams for each span. The camber diagrams take into consideration the estimated amount of camber induced in the girders when they were tensioned.

As a general rule, the primary activities are listed as column headings on the concrete take-off. Using 8½ × 11 quadrille pads having 5 squares per inch, as many as 10 columns may be used in practice. Due to lack of space in Figure 4.57, however, the intermediate diaphragm soffits and the deck edge are carried in the miscellaneous column.

In this example, no difficulty results from this manner of listing these two items, since only one entry is required for each of them. On projects involving more than one bridge, however, the totals for those sections of the individual bridge take-offs, which pertain to work that is common to more than one structure, are combined, and the combined take-off is used to set up the estimate sheets. In such cases, it is important that those primary activities common to more than one structure be listed as column headings on the take-offs for each structure, so as to facilitate the preparation of the combined take-off.

A combined take-off, together with a cost and time estimate for a project consisting of several structures, is illustrated in Section 4.19.

PRECAST CONC. GIRDER BRIDGE		GROSS		I.
	MISC.	SOF	H4	CY

DECK SLAB

ABUT. DIAPHRAGMS
  $2 \times 4.17 \times 28.25/_{27} - 5 \times 0.75 \times 4 \times 1.58/_{27}$ 7.8
  $28.25 \times 4.71 + 4 \times 4 + 4 \times 5.09 \times 4.17$ H4 234
  1 ADD'L. .... H4 234, CY 7.8
  SET 64 LF EXP. JT. ARMOR
    = 6 PCS. @ 10.7' 8.84 CWT. EA } .... MISC. 1
  FURN. & SET 40 - ¾"Ø×16" MB .... MISC. 1
BENT DIAPHRAGM
  $3 \times 4.17 \times 28.25/_{27} - 10 \times 0.75 \times 4 \times 1.58/_{27}$ 11.3
  $8 \times 5.09 \times 4.17 + 2 \times 3 \times 4$ H4 194
INT. DIAPHRAGMS
  $A = 5.09 \times 2.92 + 2.25 \times 0.5 \times 2 = 17.1$ SF
    $17.1 \times 0.667 \times 16/_{27}$ 6.8
    $17.1 \times 2 \times 16$ H4 547
  SS 16 SOF'S. @ 6.09 LF EA. .... MISC. 1
DECK
  $32 \times 0.542 \times 83.58 \times 2/_{27}$ .... SOF 5349, CY 107.4
  ADD'L @ GIR'S.
    $0.167 \times 1.58 \times 80.08 \times 10/_{27}$ 7.8
  ADD'L. @ OH $(7 + 11½) ½ = 9.25" = 0.77'$
  $(0.77 - 0.54) 1.88 \times 83.58 \times 2 \times 2/_{27}$ 5.4
  SS DE $83.58 \times 2 \times 0.583 \times 2 = 195$ SF .... MISC. 1

Totals: Σ_{26} | 4 | 5349 | 1209 | 154.3*

    * DIA'S. 33.7 CY
      DECK 120.6 CY

NEOPRENE PADS $12 \times 1½ \times 22$ 20 EA

1½" CELOTEX $(2 \times 29) 2 + 3 \times 29 = 203$
      FURNISH $7 \times 32 = 224$ SF

Figure 4.56 Concrete take-off for precast concrete girder bridge deck.

PREC'ST CONC. GIR. BRIDGE (CONT'D)	MISC.	PLY.	4x6 4x4	2. 2x8 2x6
<u>SOFFIT FORMS</u>				
X BMS. 78/6 =13,+1 = 14 SETS EA. SPAN				
14×2×4= 112 SETS 2×8 224/6				1.8
INT. JOISTS 14×2×5×4= 560				
560/5=112 2×4 112/9, 2×6 560/6				4.0
HANGERS 14×2×5 = 140 EA	1			
2×6 .SPA. 140×8/5= 224 PCS				
= 0.2 MFBM HAVE	1			
SHE-BOLTS 140 SETS	1			
½"φ RODS 56/8" 224/6" }	1			
✓ ✓ NUTS 280 EA.				
OH BRACKETS				
78/6 =13 EA SPAN, +1 @ BENT				
= 27 EA SIDE = 54 TOT.				
4×4 54/8			0.6	
2×4 54/3				0.1
MAKE 0.7 MFBM OH BR'K'S.	1			
3×3×¼ ∠ 54 PCS 4" LG-3 HOLES)				
½"φ RODS 54/40"				
✓ ✓ NUTS 108 EA }	1			
✓ ✓ CUT WASH. 54 EA				
✓ ✓ INSERTS 2 EA)				
OH JOISTS 28 SPA. @ 6' EA SIDE				
2×6 28×2×3×6 = 1008 LF				1.0
4×6 28×2×1×6 = 336 LF			0.7	
2×4 BRIDG'G. 336 LF				0.2
BLOCKING NC HAVE				
PLYFORM				
82.08×2×5.09×4 + 83.58×2×1.88×2				
= 3971 SF, X 1.17		4650		
INT. DIA. SOFFITS 2×6 32/6				0.2
	5₂₇	8	4650	1.3	7.3

Figure 4.57 Concrete take-off for precast concrete girder bridge deck (continued).

Table 4.16. Unit Costs for Precast Concrete Girder Bridge Decks

Activity	Labor		Rental		Unit of Measure
	Unit Cost	Index	Unit Cost	Index	
SS intermediate diaphragm soffits	$ 8.00	$K_{6+2} = 1000$	—		LF
Make diaphragm forms	$ 0.26	$K_{6+2} = 1000$	—		SF
SS diaphragm forms	$ 2.12	$K_{6+2} = 1000$	—		SF
Pour diaphragms	$ 7.60	$K_L = 13.48$	—		CY
Make deck overhand brackets	$450	$K_{6+2} = 1000$	—		MFBM
SS gross deck soffit forms	$ 1.51	$K_{6+2} = 1000$	—		SF[a]
SS deck edge H6″–H7″	$ 4.25	$K_{6+2} = 1000$	—		SF
Pour deck slab	$11.20	$K_{L+F} = 29.58$	$7.05	$K_{Cr} = 75$	CY
Set celotex and neoprene pads	$ 0.50	$K_{6+2} = 1000$	—		SF
Set expansion joint armor	$ 4.00	$K_{6+2} = 1000$	$4.00	$K_{Cr} = 75$	LF[b]

[a]Per SF of gross soffit area. Gross soffit area = WL, where W = overall width of deck slab and L = length of deck slab.
[b]Per LF of armored joint.

Note: The equipment cost for pouring assumes normal access for the crane, and that the concrete can be deposited directly into the forms. When pours are large or average in size, the unit costs for the crane include 1 hour of travel time. On small pours, travel time will be additional.

188

The neoprene pads as well as the area of the celotex required are noted at the bottom of sheet 1 of the concrete take-off. In practice, these materials would be included in the sections of the take-off pertaining to the abutments and the bent cap. The area of celotex noted to be furnished is based on buying the material in 4 × 8 foot sheets. The unit cost shown in Table 4.16 for placing the celotex should be applied to the total area to be furnished.

4.15.2.1 THE SOFFIT FORMWORK. The soffit formwork is based on a preliminary design, and the required materials would be taken off from sketches on the design sheets.

The rods used with the girder hangers are made up using half a she-bolt together with a ½-inch-diameter tie rod. The coarse Acme threads on the she-bolts permit a quick preliminary grade adjustment to be made from the bottom, and the precise adjustment is made with the ½-inch nut at the top of the hanger. On precast girder bridges, excessive lateral deflection of those interior longitudinal joists located at the sides of the girders must be prevented. In our case, these joists are held firmly against the girder sides with 2 × 4 bridging at 3-foot centers. If the fluid pressure of the concrete causes undue lateral deflection of these joists, leakage will occur, and the construction joint at the tops of the girders will require a considerable amount of additional but needless finishing labor.

4.15.2.2 EXPANSION JOINT ARMOR. The cost to furnish expansion joint armor is normally included in a separate job item for miscellaneous iron. The cost to install armor is included in the estimate section covering the bridge deck.

Expansion joint armor (Fig. 4.54) is furnished in two pieces, temporarily bolted together for erection. One piece consists of the two plates, and the other consists of the angle and the 1-inch-square bar. ⅜-inch-diameter open holes 18 inches on centers should be provided in both pieces to enable the armor to be firmly secured before placing the concrete.

4.15.2.3 ADDITIONAL FINISHING. The outer face of the exterior girders and the deck edge must be sacked after the slab and barrier railing have been completed. In addition, the tie rod holes in the exterior girders and in the slab soffit must be filled.

4.15.2.4 SANDBLASTING. The construction joint at the top of the girders must be sandblasted. Manufactured girders are usually sandblasted at the factory, but this must be verified in each case.

4.16   Plate Girder Bridge Decks

Plate girder bridge decks together with typical soffit formwork are shown in Figures 4.58 and 4.59. The applicable unit costs are given in Table 4.17.

The plate girders themselves are covered by a separate job item for structural steel, which is normally bid on a price-per-pound basis.

Plate girders are furnished and erected by structural steel subcontractors. Anchor bolts, when required, are furnished by the steel subcontractor but are set and dry packed by the general contractor. Steel bearing-assemblies are furnished and set by the structural steel subcontractor at the time the girders are erected. Neoprene bearing-pads are normally furnished and set by the steel subcontractor, but this must be confirmed in each case.

The cleaning and painting of plate girders is covered by a separate job item which is usually bid as a lump sum. In most instances, the girders are cleaned and primed in the shop by the structural steel subcontractor, who quotes a separate lump sum price for this portion of the painting item.

The field painting is always done by painting subcontractors. Their quoted prices will be generally proportional to the pounds of structural steel involved.

The primed steel girders will be stained at the time the deck is poured, and must be cleaned before the field painting starts. The cost of doing this work has been discussed previously in Section 4.1.

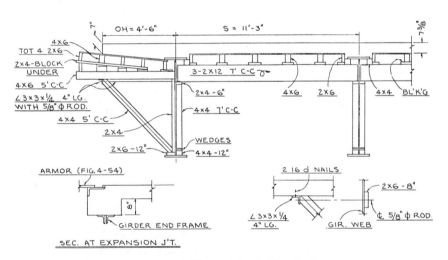

Figure 4.58 Plate girder bridge deck.

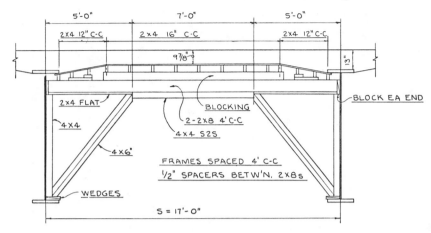

Figure 4.59 Plate girder bridge deck (continued).

4.16.1 Unit Costs

The unit costs for setting and stripping deck soffit forms are based on gross soffit areas defined in the note below Table 4.17. No deductions in area are made for the areas occupied by the girder flanges. The unit costs given include the cost to set and strip all the formwork shown, including the plyform and chamfer strips.

The unit costs shown for setting and stripping soffit forms are based on gross soffit areas ranging from 5000 to 16000 SF. For gross soffit areas of 1000 SF the unit costs for setting and stripping should be increased about 30 percent and the cost to pour the slab should be increased about 50 percent.

The unit cost for setting and stripping beam sides is applied to the vertical contact areas of deck beams such as those occurring at expansion joints.

Separate unit costs are shown for setting and stripping gross overhang soffit forms and gross interior soffit forms. In the field, these operations are performed concurrently by one crew, and it is virtually impossible to ascertain true unit costs for each operation. For the same reason, the unit costs previously discussed for setting and stripping falsework include the costs for setting and stripping that plyform immediately on the falsework.

Viewed algebraically, it would seem that separate unit costs could easily be derived by the use of simultaneous equations applied to the average unit cost data from two different jobs or decks.

Table 4.17. Unit Costs for Plate Girder Bridge Decks

Activity	Labor		Rental		Unit of Measure
	Unit Cost	Index	Unit Cost	Index	
Make deck overhang brackets	$450	$K_{6+2} = 1000$	—		MFBM
SS soffit forms S3'–S12'[a]					
Gross OH soffit	$ 2.72	$K_{6+2} = 1000$	—		SF
Gross interior soffit	$ 1.16	$K_{6+2} = 1000$	—		SF
SS soffit forms S17'[a]					
Gross OH soffit	$ 3.85	$K_{6+2} = 1000$	—		SF
Gross interior soffit	$ 1.64	$K_{6+2} = 1000$	—		SF
SS beam sides	$ 3.60	$K_{6+2} = 1000$	—		SF
SS deck edge H6″ = H7″	$ 4.25	$K_{6+2} = 1000$	—		SF
Pour deck slab[b]	$11.20	$K_{L+F} = 29.58$	$7.05	$K_{Cr} = 75$	CY

[a]When S (see note, below) varies from 3 to 12 feet, the minimum unit cost to set and strip the gross soffit area will be about $1.30 per SF. Average unit costs to set and strip gross soffit areas are:

S 3' to S 9'	$1.32 per SF
S11' to S12'	$1.44 per SF
S17'	$2.10 per SF

The preceding unit costs are indexed to $K_{6+2} = 1000$.

[b]The equipment cost assumes normal access for the crane, and that the concrete can be deposited directly into the forms. When pours are large or average in size, the unit costs for the crane include 1 hour of travel time. On small pours, travel time will be additional.

Note: S = girder spacing in feet.

Gross OH soffit area = the sum of the two OH dimensions (figure 4.58) multiplied by the length of the deck slab.

Gross interior soffit area = the width, center to center, of the exterior girders, multiplied by the length of the deck slab.

Gross soffit area = gross OH soffit area + gross interior soffit area = total area of bridge deck.

Figure 4L Plate Girder Bridge, Kern River. Contractor: Tumblin Company.

In practice, such procedures may result in derived unit costs which, although mathematically correct, are grossly different from the true unit costs. The reason for such discrepancies is that the true unit costs being realized for the separate operations are never precisely the same on different jobs, nor even from day to day on a particular job, since unit costs are influenced by many varying factors, as discussed in Chapter 2. An example clearly illustrates the problem.

Assume that two similar plate girder bridge decks have been constructed, and that the dimensions and average unit costs realized were as follows:

Figure 4M Plate Girder Bridge, Olive Drive Overcrossing. Contractor: Tumblin Company.

Figure 4N Plate Girder Bridge, Haunched Girders. Contractor: Tumblin Company.

Bridge 1
  Gross soffit area 9,000 SF
  Average unit cost to set and strip gross soffit = $1.33 per SF
  Total overhang width = 8 feet = 21.05%
  Interior soffit width = 30 feet = 78.95%
  Total width of deck = 38 feet = 100.00%
Bridge 2
  Gross soffit area 10,000 SF
  Average unit cost to set and strip gross soffit = $1.35 per SF
  Total overhang width = 7.84 feet = 20.40%
  Interior soffit width = 30.60 feet = 79.60%
  Total deck width = 38.44 feet = 100.00%.

For two similar decks, the average unit costs shown represent very close agreement, and any contractor would be pleased with such results. The above data may be restated algebraically as follows:

1. $0.7895\,I + 0.2105H = 1.33$
2. $0.7960\,I + 0.2040H = 1.35$

where I = unit cost to set and strip the gross interior soffit.

H = unit cost to set and strip the gross overhang soffit.

Solving these equations simultaneously, we find that I = $1.98 per SF and that H = $-$1.11 per SF. These unit costs are mathematically correct, but are meaningless from a practical standpoint.

When average unit cost data are available from a number of jobs, and when the contractor has an approximate idea of the magnitude of the unit cost for one of the separate activities, it is sometimes possible to derive separate unit costs for two combined activities which will closely fit the average unit costs realized on all of the jobs.

The process requires the use of good cost data, different combinations of equations, sound judgment, and some trial and error. It is only necessary or desirable to derive separate unit costs for combined activities in those cases where the separate unit costs are quite different, and where the relative amounts of each of them may vary widely on different jobs.

4.16.2   The Concrete Take-off

The concrete take-off for plate girder bridge decks is similar to the one illustrated previously for precast concrete girder decks.

Figure 40 Plate Girder Bridge, Interstate 5 Crossing U.S. 99. Contractor: Tumblin Company.

4.16.3   Soffit Forms

The formwork is based on a preliminary design. Diagonal struts are used for the 17-foot spacing, since an excessive amount of heavy timber would otherwise be required.

The deck overhang formwork for the 17-foot girder spacing will be similar to that shown in Figure 4.55, except that, when wide girder spacings are used, the spans are generally longer, the girders are deeper, and the deck overhangs are greater. In addition, cantilever construction is often used where spans are long, and the girders may be haunched at the bents.

4.17   Minor Concrete Structures

Headwalls and drainage inlets required for pipes up to about 3 feet in diameter are frequently bid under a separate job item for minor structures. The individual structures included in the item will usually contain 1.5 to 2 cubic yards of concrete each, although, when many structures are involved, a few of them may require as much as 6 or 7 cubic yards. These larger structures are frequently bid under a separate job item for structure concrete.

In most cases, the minor structures item will, in itself, constitute only a minor portion of the overall project. Thus, a simple box culvert job may contain an item for 4 cubic yards of minor concrete structures, while a freeway project may have several hundred cubic yards in the item.

The bid price per cubic yard for minor concrete structures normally includes the costs of all required excavation, backfill, reinforcing steel, and some of the miscellaneous iron, in addition to the costs for forming, placing, finishing, and curing the concrete.

Drainage inlets may require curb angles, an occasional small steel beam, or ¾-inch diameter steel rungs. The costs to furnish and install such items may be included in the price bid for minor structures. The material is furnished, F.O.B. job, at a separate lump sum price, by the miscellaneous iron supplier, and is installed by the general contractor.

The cost to furnish metal frames, covers, and grates is usually included in the job item for miscellaneous iron. This material is installed by the general contractor, and the installation cost is included in the price bid for minor structures.

Similarly, the reinforcing steel required is quoted as a separate item, F.O.B. job, by the reinforcing steel subcontractor, and this steel is installed by the general contractor.

The cost of minor concrete structures is estimated by applying overall unit costs to the total cubic yards of concrete in the item. A good set of plans will have the concrete and reinforcing steel tabulated for each structure; no additional breakdown is required, and valuable time is saved by omitting a detailed analysis of this minor job item.

The volumes of embedded pipes and of openings are deducted in determining the pay quantities for minor concrete structures.

Typical pipe headwalls and a typical drainage inlet are shown in Figure 4.60. Drainage inlets are designated by types. Some may have curb openings as shown; others may have grates, and some may require shallow collection boxes draining into the inlets themselves.

4.17.1   Unit Costs

Unit costs for minor concrete structures are shown in Table 4.18. These unit costs include the costs of excavation and backfill, the costs for installing all miscellaneous iron and reinforcing steel, and the costs to form and place the concrete complete, including all necessary finishing and curing.

In some cases, a small drywell may be required in the bottom slab of the drainage inlets. The cost to furnish and install such drywells is not included in the unit costs shown, since drywells are seldom required.

4.17.2   The Minor Concrete Structures Estimate

The estimate shown in Figure 4.61 assumes 60 CY of headwalls and 48 CY of drainage inlets are included in the item. An allowance of 8 percent is made for waste concrete since many small pours are involved and concrete is ordered in ¼-cubic yard increments.

The allowance for form lumber depends on the job item for structure concrete. In many cases materials from this item may be re-used and no additional allowance is necessary. In our estimate we assume that the plyform will be available on the job but plan to use 2-inch material from our yard. In estimating the amounts of form materials required, we use approximate factors of 20 MFBM of 2-inch material and 12 MSF of plyform per MCY of concrete involved.

The allowance for miscellaneous expenses is based on recent cost records, when available, or may be estimated if necessary. The estimated allowances for form oil, curing compound, and chamfer strip are based on a consideration of the quantities relating to one typical headwall and to one typical drainage inlet, together with the total volumes of concrete involved. The total chamfer strip required would be reduced by as much as 75 percent to allow for re-use, since short lengths are used.

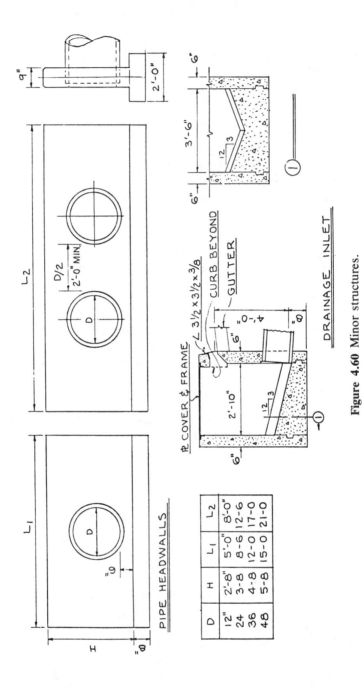

Figure 4.60 Minor structures.

D	H	L_1	L_2
12"	2'-8"	5'-0"	8'-0"
24	3-8	8-6	12-6
36	4-8	12-0	17-0
48	5-8	15-0	21-0

PIPE HEADWALLS

DRAINAGE INLET

Table 4.18. Unit Costs for Minor Concrete Structures

| | Labor | | Rental | | |
Activity	Unit Cost	Index	Unit Cost	Index	Unit of Measure
Pipe headwalls, complete	$228	$K_{6+2} = 1000$	$20	$K_{BH} = 25$	CY
Drainage inlets, complete	$336	$K_{6+2} = 1000$	$24	$K_{BH} = 25$	CY

Note: See text for discussion of the items included in these unit costs. When drainage inlets are bid as a structure concrete item, the costs will be about as follows:

SS wall forms $2.80 per SF $(K_{6+2} = 1000)$
SS soffit forms $3.00 per SF $(K_{6+2} = 1000)$
Pour concrete $40 per CY $(K_L = 13.48)$

ITEM 63 108 CY MINOR STRUCTURES			
C 16.34 K_{6+2} = 1000 L 13.48 F 16.10 K_{BH} = 25		L	DAYS	M,R

				L	DAYS	M,R
	108 X 1.08 = 117 CY		30	⎫		3510
	2X 0.108 X 20 = 2.2 MFBM HAVE		100	⎪		220
	PLY 0.108 X 12 = 1.3 MSF NC			⎪	29810	—
	MISC. EXP. 0.108 X 4200			⎪	1000x⅓	455
	REINF. FOB JOB 108 CY @ 31			⎬	=90	3350
	CURB ∠s & STEPS		LS	⎪	90/3=	600
60.0	60 CY HEADWALLS	228	20	13680	30 D	1200
48.0	48 CY DI	336	24	16130		1150
108.0				29810		10485
	INCL. PAYROLL EXP. (1.195)			35620		
	Σ = $46,105 ($ 427/CY)					

Figure 4.61 Estimate for minor structures.

Items from the general costs checklist previously listed on the structure concrete estimate are not duplicated here, except for those items whose quantities are specifically increased because of the minor structures work.

The allowance for payroll expenses is based on workmen's compensation insurance rates for concrete work NOC.

4.18   Pipe Headwalls

Headwalls for pipes over 4 feet in diameter are frequently bid under a separate job item for structure concrete. The bid unit price includes the costs to form, pour, finish, and cure the headwall complete. The attendant structure excavation, structure backfill, and reinforcing steel are paid for under separate job items for this work.

4.18.1   Unit Costs

Unit costs for large pipe headwalls are shown in Table 4.19. The cost to finish the front wall face at its junction with the pipe or pipes depends on the circumferences of the pipes and the number of pipe ends involved. The unit of measure for this unit cost is inches of pipe diameter, and this cost is in addition to the usual costs for additional finishing.

Table 4.19.   Unit Costs for Pipe Headwalls

Activity	Labor		Rental		Unit of Measure
	Unit Cost	Index	Unit Cost	Index	
Fine grade	$ 0.30	K_L = 13.48	—		SF
Make wall forms	$ 0.33	K_{6+2} = 1000	—		SF
SS foundations	$ 2.46	K_{6+2} = 1000	—		SF
SS walls H5'–H9'	$ 1.35	K_{6+2} = 1000	—		SF
Pour	$13.70	K_L = 13.48	—		CY
Finish at pipe ends	$ 0.42	K_F = 16.10	—		Inch[a]

[a]Per inch of pipe diameter, per end.
Note: When a crane is required to place the concrete, the labor cost will remain as shown, and the production rate for the crane will be about 8 to 10 CY per hour.

4.19   Combined Take-Off and Estimate

A combined take-off and estimate for a structure concrete item consisting of two flat slab bridges, three T-girder bridges and six box girder bridges is illustrated in Figures 4.62 to 4.68.

The structure excavation, structure backfill, reinforcing steel, piles, pile extensions, and bridge or barrier railings are covered by separate job items, and are not considered in the structure concrete estimate.

In practice, each structure would be taken off separately since no two structures are ever identical. For simplicity, since the required procedures remain the same, we will assume the two flat slab bridges, the three T-girder bridges and the six box girder bridges are, in each case, identical to the flat slab, T-girder, and box girder bridges whose costs have been estimated previously as single structures on separate projects.

4.19.1   The Combined Concrete Take-off

By combining the similar operations and activities from the separate structure take-offs as shown, a total of 71 separate take-off sections is reduced to 9 combined sections. This procedure simplifies the estimate sheet, facilitates the preparation of the general costs section as well as the preparation of the time of completion estimate, and greatly reduces the possibility of error.

When the combined take-off has been completed, a check must be made to ensure that all sections of each separate take-off have been listed, and that the total quantity of concrete shown on the combined take-off corresponds to the total of the amounts shown on the separate take-offs. In practice, the take-off sections for each item are numbered consecutively, and this check is easily made.

Referring to section Σ_2, for example, it will be noted that we list one section, S_{14}, and then determine the quantities for the additional identical structures by multiplication. In practice, the structures will not be identical, and the quantities for each structure will be listed. The resulting array of quantities is vital to the preparation of the general costs section, since the amount of forms to be furnished and made will generally depend on the largest quantities appearing in the array.

When space permits, repetitive items that were listed in the miscellaneous columns of the separate take-offs are listed in headed columns on the combined take-off.

The T-girder soffit area and the net slab soffit area are listed in separate columns in section Σ_5. Although their sum will be used to determine the amount of plyform required, the cost to set and strip the deck slab is based

2 FLAT SLAB + 3 T GIRDER + 6 BOX GIRDER BRIDGES

		FG	H 5/10	H6	CY H 5/10	1. CY H6
ABUTMENTS						
FLAT SLAB FIG. 4-18	S_6	256		1250		33.8
1 ADD'L. S_6		256		1250		33.8
T GIR. FIG. 4-33	S_{13}	216	1436		42.8	
2 ADD'L. S_{13}		432	2872		85.6	
BOX GIR. FIG. 4-48	S_{13}	216	1436		42.8	
5 ADD'L. S_{13}		1080	7180		214.0	
Σ_1		2456	12924	2500	385.2	67.6

			ADD'L. CONC.	NO.	FG	CY
BENT F'ND'S.						
T GIR. FIG. 4-33	S_{14}		3.5	2	200	16.7
2 ADD'L. S_{14}			7.0	4	400	33.4
BOX GIR. FIG. 4-48	S_{21}		3.5	1 L	196	18.2
5 ADD'L. S_{21}			17.5	5 L	980	91.0
Σ_2			31.5	6+6 L	1776	159.3

			NO. 6'ϕ H20	SF H20	CY O	CY □
BENT COL'S.						
T GIR. FIG. 4-33	S_{15}			560		17.8
2 ADD'L. S_{15}				1120		35.6
BOX GIR. FIG. 4-48	S_{22}		1		21.2	
5 ADD'L. S_{22}			5		106.0	
Σ_3			6	1680	127.2	53.4

		SETS BOLTS	CAP BMS.	DE H1 $\frac{25}{}$	SOF.	CY
FLAT SLAB DECK FIG 4-18	S_7	42	162	251	3456	174.2
1 ADD'L. S_7		42	162	251	3456	174.2
Σ_4		84	324	502	6912	348.4

Figure 4.62 Combined concrete take-off.

11 BRIDGES (CONT'D.)	MISC.	7"DE	GIR. SOF.	NET SLAB SOF.	H4½	CY
T GIR. DECKS FIG. 4-34 S16	1	158	885	3411	6567	265.6
2 ADD'L. S16	2	316	1770	6822	13134	531.2
Σ5	3	474	2655	10233	19701	796.8

		GIR. B'H'D.	SLAB B'H'D.	H3½	SOF.	CY
BOT. SLAB + GIR'S.						
1 BOX GIR. FIG. 4-49 S23		11	14	6994	4755	227.7
5 ADD'L. S23		55	70	34970	23775	1138.5
Σ6		66	84	41964	28530	1366.2

		7"DE	OH SOF.	LOST SOF.	CY
BOX GIR. DECKS					
1 BOX GIR. FIG. 4-49 S24		204	816	4269	121.9
5 ADD'L. S24		1020	4080	21345	609.5
Σ7		1224	4896	25614	731.4

	MISC.	PADS	CWT 2B+E	16WF36	SS H7	SS TOP	4x4 6x6	6x12	2x
FALSEWORK									
FL. SL. (4-18) S8		108			13.6		8.6	5.2	1.5
1 ADD'L.		108			13.6		8.6	5.2	1.5
T GIR. (4-34) S17	1	20	218	20/24		15.5	9.7	6.2	
2 ADD'L.	2	40	436	40/24		31.0	19.4	12.4	
BOX GIR. (4-50) S25	1	20	289	20/26		19.0	10.9	8.5	
5 ADD'L.	5	100	1445	100/26		95.0	54.5	42.5	
Σ8	9	396	2388	180	27.2	160.5	111.7	80.0	3.0

CHECK

FLAT SLABS $S_6 - S_8'$ $208.0 \times 2 = 416.0$
T GIR'S. $S_{13} - S_{18}'$ $342.9 \times 3 = 1028.7$
BOX GIR'S. $S_{13}, S_{21} - S_{25}'$ $431.8 \times 6 = 2590.8$

$$4035.5 = \Sigma_1^7 \Sigma'$$

$(S_{18} \text{ IN } \Sigma_0)'$

Figure 4.63 Combined concrete take-off (continued).

GENERAL COSTS	MISC.	MAKE	16WF36	2B-E	LOST PLY	PLY	4x4 6x6	6x12 4x12	3. 2X
Σ1-0 ABUTMENTS									
MAKE 2 S13 2872×1.17,1.75		2872				3.4			5.0
Σ3-0 BENT COL'S.									
MAKE 1 S15 560×1.17,1.75		560				0.7			1.0
100 SHE-BOLTS (4-36)									
40 ½"φ × 46" RODS }	1								
100 ✓ × 30" ✓ }									
FURN. 1 FORM 6'φ×22'	1								
Σ4-0 FLAT SLAB DECK									
MAKE DE 1 BRIDGE		251							
DE+BMS. 413×1.17,1.75						0.5			0.7
SOF. 1 BR. 3456×1.17						4.0			
Σ5-0 T GIR'S. + DECK									
MAKE 1 S16 GIR. FORMS		6567							
6567 × 1.17, 1.75						7.7			11.5
SLAB + GIR. SOF. 1 BR.									
4296 × 1.17						5.0			
S18 (FIG. 4-35) 1 BR.							1.8		4.2
Σ6-0 BOT. SLAB + GIR'S.									
(6994 × 6/5) × 1.17,1.75		8393				9.8			14.7
SOF. PLY FOR 3 BR'S.									
(4755 × 3) × 1.17						16.7			
FORM TIES 6994/8 = 9C	1								
Σ7-0 BOX GIR. DECKS									
LOST SOF. 25614 × 1.17,1.35					30.0				34.6
OH SOF. FOR 2 BR'S.									
(816×2) × 1.17,4.8						1.9			7.8
MAKE DE 1 BR.		204							
204 × 1.17, 1.75						0.2			0.4
Σ8-0 FALSEWORK									
1 FLAT SLAB							8.6	5.2	1.5
3 BOX GIR'S.				867			32.7	25.5	
WF FOR 2 BR'S.			40/26						
MAKE (108+60) PADS }	1								
= 2.7 MFBM }									
RODS (FIG. 4-40) 2 BR'S.	1								

Figure 4.64 Combined concrete take-off (continued).

GENERAL COSTS (CONT'D.)	MISC	MAKE	16WF36	2B+E	LOST PLY	PLY	4×4 6×6	6×12 4×12	2X
RE-USE FORM MAT'LS.									
Σ_{1-0} $^2/_3$× 3.4, 5.0					(2.3)				(3.3)
Σ_{3-0} 1×0.7, $^2/_3$×1.0						(0.7)			(0.7)
Σ_{4-0} 1×0.5, $^2/_3$×0.7 SOF. PLY					(0.5) (4.0)				(0.5)
Σ_{5-0} SLAB+GIR. SOF. GIR. FORMS					(5.0) (7.7)				(11.5)
Σ_{6-0} GIR FORMS AVAILABLE 6994× $^4/_5$ ×1.17, 1.75 = 6.5 & 9.8					(2.3)				(5.8)
Σ_{8-0} FLAT SLAB									(1.5)
Σ_0	5	18847	40/26	867	8.2	49.2	43.1	30.7	58.1
HAVE					8.2	20.0	30.0	20.0	9.0

Figure 4.65 Combined concrete take-off (continued).

on the net slab soffit area, while the cost to set and strip the plyform for the girder soffits is included in the cost to set and strip the falsework on which it rests.

In section Σ_8, the 1.7 MFBM of pad material in section S_8 is included in the column for 4×4 material. Although 2×12 material will be used for the pads, we plan to furnish both materials from our yard, and the charge will be the same.

4.19.1.1 GENERAL COSTS SECTION Σ_0.

The take-off for the general costs section is made directly from the combined sections Σ_1 through Σ_8.

At this time the operating portion (sections Σ_1 through Σ_8) of the estimate has not been made, and our decisions concerning the amounts of forms to be furnished are based on judgment and experience. When the operating portion of the estimate has been priced, the time required for completion will be determined, and these decisions will be reviewed and modified if necessary.

Since the flat slab bridges are small, and since they are the only ones requiring wood falsework, we plan to construct them concurrently with the first four T- and box girder bridges. This will keep the flat slab bridges off the critical path, and will permit us to re-use the flat slab forms in the lost soffit forms of the box girder bridges.

We plan to start one crew working on the abutments, and to start a second crew working on the bent foundations and columns. These crews will work full-time on these operations until the abutments and bents have been completed.

In addition, we will have one crew setting and stripping falsework, and one or two crews working on girders and slabs. To keep these crews working steadily, falsework will be furnished for one flat slab bridge and for three box girder bridges, except that wide flange beams need be furnished for only two structures.

At any given time we will be working on three box or T-girder bridges which will be constructed in the following order:

Group 1. T-girder
                Box girder
                T-girder
Group 2. Box girder
                T-girder
                Box girder
Group 3. Box girder
                Box girder
                Box girder

This sequence assumes that job conditions do not limit the order of work. 71 CWT more steel shoring must be furnished than would be required if each group consisted of one T- and two box girder bridges. The order selected above will, however, permit all of the T-girder forms to be re-used in the lost soffit forms of the group 3 structures.

Section Σ_1 indicates that two pairs of S_6 abutments and nine pairs of S_{13} abutments must be formed. To eliminate excessive costs for additional finishing, we prefer to limit the number of re-uses of forms for most exposed surfaces to about six. Therefore, two sets of abutment forms must be furnished. We will actually make one set of S_6 forms and one set of S_{13} forms, but will modify the S_6 forms for re-use. For estimating purposes, we make two sets of S_{13} forms in section $\Sigma_{1\text{-}0}$, and assume the cost to modify the forms is included.

No form material will be required for section Σ_2, since we plan to pour the footings neat. The reinforcing steel templates and the starter wall forms for the rectangular columns will be made from material furnished for the other work. No starter walls will be used for the circular columns.

One set of S_{15} column forms is made in section $\Sigma_{3\text{-}0}$, and we note that one steel column form will be required. The she-bolts and tie rods are listed, and an allowance is made for the one 30-inch rod that will be embedded and lost at each set of wales.

In section $\Sigma_{5\text{-}0}$ we plan to furnish form material for one complete T-girder bridge. The deck edge forms are not furnished in section $\Sigma_{5\text{-}0}$, since we note that a greater quantity of identical forms are required in section Σ_7. In this case, 9 re-uses is satisfactory, since no form ties or reinforcing dowels pass through these forms, and the fluid pressure of the concrete on them and their related rate of wear are small.

In section $\Sigma_{6\text{-}0}$ we plan to furnish bottom soffit plyform for three box girder bridges, girder forms for one complete box girder bridge, and an additional set of forms for the outside faces of the exterior girders. The additional forms are necessary because the exterior forms must remain in place until the deck slabs are poured and cured.

The form ties are noted in the miscellaneous column, and these ties will suffice for the entire job. We will use taper ties, which are completely reuseable and have no embedded or lost parts.

In section $\Sigma_{7\text{-}0}$ we plan to furnish deck overhang soffit forms for two box girder bridges. All the lumber required is listed in the 2X column because space is limited. We plan to furnish some 2X material and 4X material from our yard, and the charges are about the same for both.

The required falsework is listed in section $\Sigma_{8\text{-}0}$. Flat pads are required for one flat slab bridge and for three sets of box girder end bents. We assume these pads must be made up for the job. The flat pads for the steel shoring towers are assumed to be made up and available with the steel shoring in our yard. The deductions for material re-use are made below section Σ_{8-0}.

The material from Σ_{1-0} will be re-used in the lost soffit forms. The material from Σ_{3-0} will be re-used in the girder forms for the box girders. The material from Σ_{4-0} will be re-used in the lost soffit for the second box girder bridge in group 2. The soffit plyform for the flat slab bridges will have only one re-use as soffit formwork; if it were new material, we would consider it too valuable to use in the lost soffit forms. In this case we plan to form the flat slab soffits with reasonably good, used plyform available in our yard, and this material will then be re-used in the lost soffit forms.

The form materials from the T-girder bridge will be available for re-use in the lost soffit forms for the last three box girder bridges, and the interior girder forms used on the last box girder bridge will be available for re-use as needed in the lost soffit forms for the last box girder.

Below the totals for section Σ_0, we note the quantities of the form materials that are available from our yard.

4.19.2 The Structure Concrete Estimate

The structure concrete estimate is shown in Figures 4.66 to 4.68. The unit costs used for sections Σ_1 through Σ_8 are identical to those used for the single structure estimates, and are shown in Tables 4.5, 4.6, 4.9, 4.10, 4.11, 4.13 and 4.14 and in Figures 4.16 and 4.26.

The item for minor expenses is taken from recent cost records when these are available. In our case, we use a weighted average of the costs determined in our separate structure estimates.

The cost for additional curing over weekends will vary from job to job. On smaller projects, it is frequently possible to have the cure handled on a part-time basis by a workman living near the job. On larger jobs, the weekend cure will require continuous attention on those occasions when large pours are being cured. Based on $K_L = 13.48$, and assuming that time and one-half is paid for weekend work, the labor cost shown represents a reasonable order of magnitude for this expense.

When a water supply is available near each bridge site, the material and equipment costs for additional curing are small. On many jobs, however, water trucks may be necessary. The rental cost shown in the estimate assumes the part-time use of a water truck on the weekend cure.

The cost used in the structure concrete estimate to develop water will also vary widely with the actual job conditions. On this project we assume that there is a large job item for earthwork, that one or more water trucks

ITEM 57 4040 CY STRUCTURE CONCRETE

C 16.34 K_{6+2} 1000
L 13.48 K_{Cr} 75
F 16.10

		L	DAYS	M,R
Σ_0 GENERAL COSTS				
CONC. 4036×1.02=4117 CY	30			123510
49.1 MFBM 2X BUY	300			14730
9.0 " " HAVE	100			900
10.7 " 6×12 BUY	360			3850
20.0 " " HAVE	120			2400
13.1 " 4×4,6 BUY	350			4580
30.0 " " HAVE	120			3600
29.2 MSF PLY BUY	500			14600
20.0 " " HAVE	170			3400
8.2 " LOST PLY HAVE	100			820
867 CWT 2B+E 8 MOS.	1734			13870
374 CWT 16WF36 4.5 USES	1.50			2520
1 COL. FORM 6'φ×22' 2 MOS.	660			1320
9C TAPER TIES 8 MOS.	40/C			2880
1C SHE-BOLTS 1 "	"			40
40 ½"φ×46" RODS	1.20			50
100 " ×30" "	0.80			80
MINOR EXP. 4.04 MCY	2740			11070
MAKE 18850 SF	0.26	4900		—
ADD'L. FIN. 4.04 MCY	3930	15880		635
" CURE 35 WKS. 225	125	7880		4380
SAND BLAST 4.04 MCY	660	2670		445
COMPRESSOR 8 MOS.	400			3200
VIBRATORS "	200			1600
TABLE SAW "	120			960
DECK FIN. MACH. 55200 SF	0.05			2760
GENERATORS (3½) 8 MOS.	875			7000
DEVELOP WATER				2000
HOPPERS				50
CURE RUGS				600
MISC. IRON				1320
BOX GIR. DOBIES 1.37 MCY	700	960		60
SWEEP DECKS 55.2 MSF	13	720		—
CLEAN UP 124.7 MFBM 43	12	5360		1500
62.1 T 34	7	2110		435

Figure 4.66 Estimate for combined structures.

ITEM 57 4040 CY STR. CONC. (CONT'D.)

				L	DAYS	M,R
	Σ₁ ABUTMENTS					
	FG 2456 SF		0.40	980	29960	
	SS 2500 SF H6		1.51	3780	1000	
	SS 12924 SF H5/10		1.95	25200	=30 D	
67.6	P 69 CY H6		4.35	300		
385.2	P 393 CY H5/10		5.25	2060		
	Σ₂ BENT F'ND'S.					
	31.5 CY ADD'L. CONC.	1	30	30	1695	945
	FG 1776 SF		0.28	495	1000x1/3	
	SS 6 EA		93	560	=5 D	
	SS 6L EA		1.10×93	610		
159.3	P 162 CY		4.45	720		
	Σ₃ BENT COL'S.					
	SS 1680 SF H20 □		1.84	3090	5250	
	SS 6 EA 6'φ×H20	360	150	2160	1000x2/3	900
	CRANE ON-OFF ADD'L. 7HRS		75	—	=8 D	525
53.4	P 55 CY	8.30	10.50	455		580
127.2	P 130 CY	8.30	10.50	1080		1370
	Σ₄ FLAT SLAB DECK					
	SET 84 SETS POST BOLTS		11	925	2595	
	SS 324 SF CAP BMS		0.87	280	1000x2/3	
	SS 502 SF DE H15"		2.76	1390	=4 D	
348.4	P 355 CY	8.54	4.40	3030		1560
	Σ₅ T GIR + SLAB					
	SS 54 SF DECK B'H'D		5.30	285	51950	
	SS 78 SF GIR ✓		1.89	145	1000x4/3	
	SS 19701 SF GIR H4²		1.89	37230	=39 D	
	SS 10233 SF SOF.		1.20	12280		
	SS 474 SF DE H7"		4.25	2010		
796.8	P 813 CY	7	5	5690		4060
	Σ₆ BOT. SLAB + GIR'S.					
	SS 84 SF DECK B'H'D.		5.30	445	48780	
	SS 66 SF GIR ✓		1.15	75	1000	
	SS 41964 SF GIR H3²		1.15	48260	=49 D	

Figure 4.67 Estimate for combined structures (continued).

				L	DAYS	M,R

ITEM 57 4040 CY STR. CONC. (CONT'D.)

				L	DAYS	M,R
13662	P 1394 CY	7.75	5.80	10800		8090
	Σ_7 BOX GIR DECKS					
	SS 30510 SF (LOST+OH) SOF.		1.13	34480	39680	
	SS 1224 SF DE H7"		4.25	5200	1000	
731.4	P 746 CY	11.20	7.05	8360	≈40D	5260
4035.5	Σ_8 FALSEWORK					
	WF RODS IN MISC. IRON					
	MAKE 2.7 MFBM PADS		115	310	10110	
	SS 396 EA FLAT PADS		7	2770	1000	
	SS 27.2 MFBM H$_c$7		270	7340	=10D	
	SS 2388 CWT 2B+E	8.90	0.60	21250	62320	1430
	SS 16WF36 180/26	40	29	7200	1000	5220
	SS 160.5 MFBM (TOP)	211	54	33870	=63 D	8670
				325625		269775
	INCL. PAYROLL EXP. (1.315)			428195		

$\Sigma = \$697,970$ (173 $/CY)

TIME EST.	CONSEC.	CONC'T
ON-OFF	5	5
ABUTMENTS	30	10
BENT F'ND'S.	5	—
BENT COL'S.	8	5
FLAT SLAB DECK	4	—
T GIR + SLAB	39	39
BOT. SLAB + GIR	49	49
BOX GIR DECK	40	40
FLAT SLAB F'W'K.	10	—
T & BOX GIR ✓	63	15
REINF. STEEL	5	5
CURE LAST BOX	10	10
	268 D	178 D
		= 8.9 MOS

Figure 4.68 Estimate for combined structures (continued).

will be in use on the job, and that there is a separate job item for developing the water supply.

The allowance for developing water in our structure concrete estimate covers the costs of setting and moving small contractor-owned storage tanks, the rental of a 2-inch pump, and the purchase of garden hose and nozzles. As an alternative, an old 2000-gallon water truck is frequently used to supply water for the structure concrete work.

4.19.2.1 THE TIME REQUIRED FOR COMPLETION. The number of working days required to complete each section is determined in the "days" column of the estimate sheet. We assume that six-carpenter crews will work on sections Σ_1, Σ_6, Σ_7 and Σ_8, a two-carpenter crew on section Σ_2, four-carpenter crews on sections Σ_3 and Σ_4, and an eight-carpenter crew on section Σ_5.

The required working days are summarized at the bottom of the sheet, and 9 months will be required to complete the item. Since fifteen working days are allowed for curing and moving on and off the job, 8-month rental periods are used in the general costs section.

4.19.3 Reduced Costs on Multi-Structure Projects

The savings realized when several structures are combined on a single project are clearly shown in the following tabulation, which is based on the preceding estimates.

	Total Cost $/CY	General Costs Bare Labor $/CY	M, R $/CY
1 Flat Slab	133	8.75	57.60
1 T-Girder	235	21.20	82.90
1 Box Girder	194	15.80	73.20
11 Structures			
Weighted Average	198	16.40	74.00
Combined Estimate	173	10.00	57.30

Since identical unit costs are used in the operating sections of all the estimates, it is clear that all savings are made in the general costs section of the combined estimate. It is not possible to estimate closely the general costs section of a structure concrete estimate without resorting to the detailed analysis illustrated herein.

The estimator should record the unit costs, in dollars per cubic yard, for each section of each estimate he prepares. As succeeding estimates are made, the costs determined should be compared to these recorded costs, and this procedure will sometimes enable errors to be detected.

4.20 Miscellaneous Concrete Costs

Unit costs for a number of miscellaneous activities encountered in the construction of engineering structures are given below. These unit costs may also be useful in estimating the cost of similar work by the comparison method.

4.20.1 Post Inserts

Indexed to K_{6+2} = 1000, 3-inch diameter by 12-inch-long galvanized cans or pipe post inserts were installed at an average unit cost for labor of $6.90 each.

4.20.2 Rubber Water Stop

Indexed to K_{6+2} = 1000, 6″ rubber water stop was installed at expansion joints at a unit cost for labor of $0.77 per linear foot.

4.20.3 Poured-in-place Falsework Footings

92 footings, 20 inches square by 8 inches deep, were excavated in dry sand and poured at a unit labor cost of $7.00 each or $102 per cubic yard, indexed to K_L = 13.48.

620 footings, 30 inches square by 15 inches deep, were excavated in dry sand and poured at a unit labor cost of $12.80 each or $44 per cubic yard, indexed to K_L = 13.48.

4.20.4 Wall Blockouts

Blockouts, 2 feet square, were placed in a form for a 12-inch wall at a unit labor cost, indexed to K_{6+2} = 1000, of $67 each or $8.33 per square foot of blockout form.

Concrete pipes, 18 inches in diameter, were inserted in the blocked out openings and were then formed and poured into the wall at a unit labor cost, indexed to K_{6+2} = 1000, of $70 each.

Blockouts, 18 inches square, were placed in retaining wall forms 15 inches thick at a unit labor cost, indexed to K_{6+2} = 1000, of $64 each or $8.53 per square foot of blockout form. The blocked out openings were later formed and poured at a unit labor cost, indexed to K_L = 13.48, of $20 each.

4.20.5   Bridge Curbs

Indexed to K_{L+F} = 29.58, 35 CY of bridge curb 2 feet wide and 9 inches high were poured, given a light trowel finish, and water cured at a unit labor cost of $24.50 per cubic yard. 100 CY of bridge curb 2 feet wide and 15 inches thick were poured, finished and cured on the same job at a unit cost for labor of $23.00 per cubic yard.

4.20.6   Slope Paving

Indexed to K_L = 13.48, 1-½:1 slopes with a vertical height of 12 feet were

fine graded for a unit labor cost of $0.28 per square foot, and $\dfrac{6 \times 6}{10 \times 10}$

mesh was placed on the slope at a unit labor cost of $0.12 per square foot.

Indexed to K_{L+F} = 29.58, a 4-inch-thick concrete slab was poured on the slope, finished, and cured with curing compound, at a unit cost for labor of $0.51 per square foot. The slab was floated with wood floats and given a light broom finish.

4.20.7   Concrete Posts

248 posts, 6 inches by 9 inches in cross section by 2 feet 6 inches in height, were set and stripped at a unit labor cost, indexed to K_{6+2} = 1000, of $10.50 each or $4.33 per square foot. Indexed to K_L = 13.48, the unit labor cost to pour the posts was $7.10 each or $150 per cubic yard.

4.20.8   Pour Sign Foundations

Indexed to K_L = 13.48, 12 cubic yard sign foundations were poured directly from the transit-mix truck and water cured at a unit cost for labor of $3.10 per cubic yard.

4.20.9   Pour With Conveyor

Indexed to K_L = 13.48, bridge bent columns and caps were poured with a portable conveyor at a unit labor cost of $11.00 per cubic yard.

4.20.10   Precast Flat Slab Deck Units

Indexed to K_{L+F} = 29.58, a casting bed consisting of 2200 SF of concrete, 4 feet wide by 4 inches thick, was fine graded, formed, poured, finished, and cured at a unit cost for labor of $0.63 per square foot.

Indexed to $K_{6+2} = 1000$, 4100 SF of side forms 1.67 feet in height, were set and stripped at a unit cost for labor of $1.68 per square foot, and 216 pcs of treated cardboard void forms, 10 inches in diameter by 13 feet in length, were installed and tied down at a unit cost for labor of $6.90 each or $0.53 per linear foot.

Indexed to $K_{L+F} = 29.58$, 218 CY of slab units 4 feet wide by 30 feet in length were poured, finished, and water cured at a unit cost for labor of $11.80 per cubic yard.

The units were erected with a 2-inch space between units, and this space was filled with grout. Indexed to $K_{6+2} = 1000$, 1000 linear feet of 2-inch-wide grout soffit was set and stripped at a unit cost for labor of $2.56 per linear foot. The soffits of the erected deck units were approximately 25 feet above the ground surface.

Indexed to $K_L = 13.48$, 11 CY of grout was poured, finished, and cured, at a unit cost for labor of $38 per cubic yard. The grout soffit forms were held with she-bolts at 5-foot centers, and these holes were filled, on completion, at a unit cost for labor of $5.20 each, indexed to $K_L = 13.48$.

4.20.11 Seal Form Panels

Indexed to $K_L = 13.48$, large quantities of plywood form panels, 8 feet square, were sprayed with form sealer, using a garden-type spray can, at a unit cost for labor of $0.03 per square foot.

4.20.12 Sheet Metal Radius Forms

The ends of the interior walls of a box culvert may be semicircular in cross section and, in such cases, the ends of the exterior walls are formed in the shape of a quarter circle. Galvanized sheet metal is used for this work, and, indexed to $K_{6+2} = 1000$, the sheet metal will be set and stripped at a unit cost of $2.80 per linear foot.

4.20.13 Stripping Forms

Unit costs for formwork are based on the total cost to set and strip the forms. The unit cost to strip wall forms will average about 25 percent of the total setting and stripping cost.

5

Structure Concrete—Buildings

Buildings are bid on a lump sum basis, and there is no item for structure concrete as such. A section in the specifications will cover the concrete work, but there is no engineer's estimate of the quantity involved.

The procedures to be followed in taking off and estimating the costs of concrete in buildings are identical to those illustrated in Chapter 4 for engineering structures.

As a general rule, the unit costs for structure concrete in buildings are higher than those for similar operations on engineering projects; buildings are more irregular in plan and in elevation, their sections are not as massive, and access for men and for equipment is often limited.

In this chapter, typical components of concrete buildings are illustrated, the quantities to be taken off are discussed, and typical unit costs are presented. The unit costs realized on building work tend to show more variation, from job to job, than do the unit costs for engineering structures.

Engineering structures are designed primarily for utility and economy, and are standardized to a great extent. Architects, however, strive to make each building different from their preceding designs, and to create, at a reasonable cost, a structure that is both beautiful and functional. Each building, therefore, tends to reflect the individual tastes and preferences of its designer, and each, to a varying degree, presents the contractor with a unique set of problems and required procedures.

The ability to modify the unit costs developed on past work to fit the particular conditions on a proposed project requires sound judgment and a solid background of field and office experience, since the estimator must

Figure 5A Bakersfield Memorial Stadium, Bakersfield College. Contractor: Tumblin Company.

be thoroughly familiar with the job conditions relating to those unit costs being adjusted.

5.1   General Costs

The discussion and the general costs checklist presented at the beginning of Chapter 4 also apply to structure concrete estimates for building work.

5.1.1   Screeds

Screeds are required for all slab pours in building work, and the rental charge for the screed supports and brackets is a general costs item. The labor cost to set screeds is included in the cost to pour the slab.

5.1.2   Additional Finishing

The costs for additional finishing on building work are based on a take-off of the quantities of work required. The estimator must, therefore, be thoroughly familiar with the requirements of the specifications and with the interior finish schedules for all rooms having concrete walls or ceilings.

5.1.3 Sandblasting

Sandblasting construction joints is seldom required on building work. Laitance is water-cut from the construction joints, the dowels are washed clean after each pour, and the costs of this work are included in the unit costs to place the concrete. When sandblasting is required, the areas involved are included in the concrete take-off.

5.1.4 Hoisting and Moving Concrete

The costs to move concrete vertically and horizontally from the point of delivery to the pour location represent a significant portion of the concrete placing costs on building work.

5.1.4.1 CONCRETE BUGGIES. Man-powered concrete buggies or carts are frequently used to deliver concrete laterally to the pour location. Typical buggies have a water level capacity of 6 CF and will deliver about 4.5 CF of concrete per load.

Unit costs for buggying concrete are shown in Table 5.1. To achieve these unit costs, the buggies should be loaded from hoppers and must operate in continuous cycles on level runways. Loaded buggies deliver concrete on one runway and return empty on another. When flat ramps in the runway pattern are unavoidable, an additional laborer must be stationed at the ramp to assist the loaded buggies up or down.

Buggy runways are made in sections 4 feet wide by 8 feet long, using 1×6 boards on 2×4 runners. These sections must usually be supported on wooden horses or light wooden bents spaced on 4-foot centers. Based on $K_{6+2} = 1000$, the runways will cost about \$0.30 per SF to make, and will require 1.75 MFBM of material per MSF. The cost to make the horses will be about \$300 per MFBM.

When buggies are required, the number of buggies together with the estimated amount of runways to be furnished are noted in the miscellaneous column of the take-off for the section requiring them, and the cost for furnishing these items is carried in the general costs section of the concrete estimate. The costs to set the runways and to move them as required during the pour are included in the unit costs to pour the concrete.

5.1.4.2 HOISTING CONCRETE. Tower or jumping cranes are used to hoist all materials on tall multi-story buildings. Ideally, the crane or cranes will cover the entire building area, all walls and floors are poured directly with concrete buckets, and no buggies are required.

The entire cost of tower cranes is a direct job overhead expense, as

Table 5.1. Unit Costs for Building Foundations

Activity	Labor Unit Cost	Index	Unit of Measure
FG spread and wall footings	$0.28	$K_L = 13.48$	SF
FG grade beams	$0.32	$K_L = 13.48$	SF
Make forms for foundation walls and grade beams	$0.32	$K_{6+2} = 1000$	SF
SS wall footings H1[a]	$2.35	$K_{6+2} = 1000$	SF
SS foundation walls H2–H4	$2.10	$K_{6+2} = 1000$	SF
SS spread footings and pedestals[a]	$2.25	$K_{6+2} = 1000$	SF
SS grade beam tops only[a]	$4.75	$K_{6+2} = 1000$	SF
SS grade beams H2–H3	$1.35–$ 1.90	$K_{6+2} = 1000$	SF
Pour foundations			
Directly from truck[b]			
Wall footings and foundation walls	$8.00	$K_L = 13.48$	CY
Spread footings	$7.00–$10.00	$K_L = 13.48$	CY
Grade beams	$4.75–$ 9.00	$K_L = 13.48$	CY
16″ Φ CIDH piles	$4.00–$ 6.00	$K_L = 13.48$	CY
Additional to set and move buggy runways	$5.00–$ 9.00	$K_{6+2} = 1000$	CY
Additional to buggy concrete[c]			
Up to 50 feet one way	$3.75	$K_L = 13.48$	CY
100 feet one way	$4.50	$K_L = 13.48$	CY
150 feet one way	$5.25	$K_L = 13.48$	CY
200 feet one way	$6.00	$K_L = 13.48$	CY

[a]The cost to make the forms is included.

[b]If the foundations are poured with a crane, the production rates will be about as follows:

Wall footings and foundation walls	10 CY/hr
Spread footings	10–15 CY/hr
Grade beams	7–15 CY/hr
Foundation pads containing 20 CY or more of concrete	25–30 CY/hr

See section 5.1.4.3 for discussion of pour costs when concrete pumps are used.
[c]Unit costs assume hand-powered buggies operating under average conditions, that the full buggy production can be utilized, and that unneeded buggies are taken off the pour as the haul distance decreases.

discussed in Chapter 3, and no additional charges are necessary when they are used to place concrete.

Concrete hoists together with buggy hoppers at the pour elevation are often used on buildings two to five stories high. These hoists are equipped with self-dumping concrete buckets and are not suitable for use as general material or personnel hoists. Concrete hoists are normally rented on a monthly basis for the duration of the elevated concrete work. The rental cost, the on and off charges, the costs to set up and to dismantle the hoist, to add tower sections as the work progresses, and the costs of fuel, maintenance, and the operator are all included in the general costs section of the concrete estimate.

The labor cost for the hoist operator may be based on an average pour rate of 6.5 CY per hour for wall concrete and 12 CY per hour for floor pours, which will account for the full day's pay that is required on part-day pours. The operator will only be hired on those days that the hoist will be used. The remaining costs noted in the above paragraph vary with the particular hoist being used, and the amount allowed is based on conversations with the supplier or manufacturer, and on the estimator's judgment.

Some concrete hoists are portable, and may be towed to and from the job; on work over one story high, however, they are not easily moved at the site, so the hoist is set up at a central location and the concrete is buggied to the pour.

On buildings up to two stories high, truck cranes are often used to hoist concrete, and little or no buggying is necessary.

5.1.4.3 PORTABLE CONCRETE PUMPS. In many areas, truck-mounted concrete pumps are now available. These pumps are equipped with hydraulic booms, and can pump concrete to heights of 100 feet or more. The concrete is delivered to the pumps by transit mix trucks. Rubber hose, attached to the boom, enables concrete to be placed directly into wall or column forms, when accessible, and no collection hoppers or elephant trunks are required. Additional sections of hose are added when the concrete must be moved laterally to areas that are inaccessible to the pump's boom. The rate of pouring will be governed by the job conditions, since typical concrete pumps have ample capacity.

The equipment cost for pumping concrete is based on an hourly rental rate for the equipment, and the rental rate for a concrete pump equipped with a hydraulic boom is currently about equal to the rental rate for a 35-ton truck crane. Since the rate of pour is governed by the job conditions, the equipment cost to pour with a pump is about the same as when a truck crane is used.

Figure 5B Truck-mounted concrete pump, A. & H. concrete pumping. Bakersfield, California.

When the pour location is accessible to the pump's boom, the labor cost to pour concrete walls, columns, or slabs will be essentially the same as when the concrete is placed with a crane, or is poured directly from transit mix trucks.

When the pour location is not accessible to the pump's boom, additional labor will be required to handle the hose, and the amount allowed will depend on the anticipated job conditions, and on the estimator's judgment. Assuming job conditions are such that two laborers will be able to handle the hose, the labor cost to pour columns in such locations will be about 25 percent greater than to pour them directly with a crane, and the labor cost to pour walls will be about 30 percent greater. The labor cost to pour slabs or floors, under such conditions, will be about 20 percent greater than to pour them directly from transit mix trucks or with a crane.

When concrete must be moved laterally to the pour location, the concrete pump will eliminate the need to buggy concrete and handle runways. In every case, the resulting savings will exceed the added costs of handling the hose.

5.2 The Concrete Take-off

The concrete take-offs for building work are identical to those illustrated in Chapter 4 for engineering structures. In all cases, the concrete take-off follows the construction sequence.

Thus, section 1 of the take-off will cover the foundations, section 2 will cover the basement walls, section 3 will cover the basement columns, and section 4 will cover the first floor slab. This process is repeated, floor by floor, until all the concrete in the walls, columns, floors, and roof has been taken off.

The concrete stairs are then processed in a separate section in which each stair is taken off, from the bottom to the top. The interior and exterior grade slabs are taken off next, and the miscellaneous exterior concrete work is processed last.

On completion, the totals for those take-off sections that relate to similar types of work are collected and summarized in a combined take-off. The process is identical to that illustrated in Chapter 4, where a combined take-off for several bridges was presented.

Structure excavation and backfill are taken off as a separate operation, after the concrete and in many cases the carpentry work, have been processed. Critical dimensions and elevations, determined in connection with the concrete take-off, are penciled in on the foundation plan to facilitate the structure excavation and backfill take-off.

5.3 Foundations

A portion of a typical foundation plan for a concrete building is shown in Figure 5.1, together with a section through a typical grade beam. Grade beams are often used in the foundations for buildings, although none are required in the foundation plan illustrated.

In practice, the footings, pedestals, pilasters, and all columns, floor by floor, are usually detailed in a single table on the structural plans. Reinforcing steel and the dimensions of the various sections are indicated by

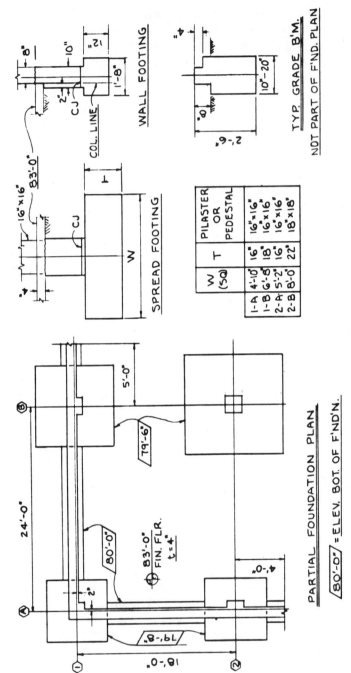

Figure 5.1 Building foundations.

letter designations on typical sections, and the actual bar sizes and spacings, together with the dimensions of the structural sections themselves, are presented in the tabulation.

5.3.1 The Foundation Concrete Take-Off

The concrete take-off is similar in all respects to those illustrated in Chapter 4.

Footing 2-B is taken off as follows:

Fine grade	8.00 × 8.00	=	64 SF
Footing forms H1.8	8.00 × 1.83 × 4	=	59 SF
Footing concrete	8.00 × 8.00 × 1.83/27	=	4.34 CY
Pedestal forms H1.3	1.50 × 4 × 1.33	=	8 SF
Pedestal concrete	1.50 × 1.50 × 1.33/27	=	0.11 CY

In taking off footing 2-A, no deduction in form area is made at the intersection of the spread footing and the wall footings, and the take-off will be as follows:

Fine grade	5.17 × 5.17	=	27 SF
Forms H1.3	5.17 × 1.33 × 4	=	27 SF
Concrete	5.17 × 5.17 × 1.33/27 + 0.33 × 1.67 × 0.67 × 2/27	=	1.34 CY

The concrete calculation includes an allowance for that additional concrete required where the wall footings will be stepped down 4 inches at their intersection with the spread footing. By including this small amount of concrete here, it will not be necessary to refer back to the spread footing depth when the wall footings are taken off.

The take-off for that length of wall footing shown in Figure 5.1 is as follows:

Fine grade	1.67 × 34.33	=	58 SF
Forms H1	1.00 × 2 × 34.33	=	69 SF
Concrete	1.67 × 1.00 × 34.33/27	=	2.12 CY

In taking off concrete walls, a vertical section together with the related plan length of wall are used. Where corners are involved, the wall lengths are taken to the center lines of the intersecting walls. This procedure allows the single wall length determined to be used to calculate both the contact form area of the wall and the CY of concrete in it. The additional

Figure 5C Science Building, California State University, Fresno. Contractor: Tumblin Company.

areas and concrete due to pilasters are based on separate calculations in each case.

The take-off for the foundation wall shown in Figure 5.1 is as follows:

Wall forms H2	2.00 × 2 × 51.50	
	+ 1.00 × 2 × 2	= 210 SF
Concrete	0.833 × 2.00 × 51.50/27	
	+ 0.50 × 1.33 × 2.00/27 × 2	
	+ 0.50 × 0.50 × 2.00/27	= 3.29 CY

In taking off the required form lumber, an allowance should be made for the additional material required to form the slab key at the top of the foundation wall, since this will be a separate piece nailed inside the wall form. The cost to set and strip all solid keyways is included in the unit costs to set and strip the walls.

In the preceding take-offs, no allowance is made for pour bulkheads; their number is not known at this time, they constitute only a small portion of the work to be done, and their costs are included in the unit costs for the related formwork.

No allowance is made for forming the starter walls, and they are not considered when determining concrete volumes. The costs to set and strip starter walls are included in the unit costs to set and strip the related formwork, and the volumes of concrete involved are small.

Grade beams may be poured in neat excavations as shown, or they may be completely formed. The grade beam illustrated will require 1.33 SF of grade beam top to be set and stripped per linear foot of grade beam. If the grade beam were to be completely formed, 5.00 SF of grade beam forms would be required per linear foot, and the unit cost used for setting and stripping would include the cost of forming the top portion of the grade beam.

When grade beams are poured neat, the specifications may require increasing the section 2 to 4 inches in width. The additional concrete required by such specifications and by over-width excavation is included directly in the concrete take-off. Such concrete is carried separately on engineering work to facilitate the checking of the engineer's estimated volume of concrete.

5.3.2 Unit Costs

Unit costs for building foundations are shown in Table 5.1.

The unit cost to set and strip grade beams will be low when long, unobstructed lengths are involved. When the grade beams intersect with spread footings or when short lengths with many corners are required, higher unit costs apply.

The unit costs for placing concrete include the costs of preparing for the pour, cleaning up afterwards, curing during regular working hours, and placing the concrete itself.

Figure 5D Kern County Jail, Bakersfield, California. Contractor: Tumblin Company.

The low unit cost for pouring grade beams applies to wide grade beams having little or no top restrictions. The high unit costs apply to narrow grade beams with severe top restrictions.

A meaningful analysis of the cost to set and move buggy runways for foundation pours cannot be made at the time the estimate is being prepared because too many assumptions would be required. The unit costs shown do represent an order of magnitude for this work, but sound judgment, based on experience and past job cost records, must be used in assigning the probable cost for this item.

5.4   Walls

A partial floor plan of a beam and slab floor is illustrated in Figure 5.2, together with a typical basement wall section.

5.4.1   The Concrete Take-off

Building walls are taken off one floor at a time, and the height of the walls depends on the floor system used for the floor above. As a general rule, the construction joint at the top of the wall is located at the soffit elevation of the deepest beam framing into it, and the bottom construction joint is at the floor line.

A vertical wall section is used to calculate the volume of the wall concrete and the contact area of the wall forms. The length used for both calculations is taken to the intersection of the wall centerlines at corners. The additional volumes and areas relating to pilasters are calculated separately.

The take-off for the wall shown from column line B to column line 2 is as follows:

Basement wall forms H10

$$9.83 \times 2 \times 42.67 + 0.67 \times 2 \times 9.83 \qquad = 852 \text{ SF}$$

Concrete

$$0.667 \times 9.83 \times 42.67/27$$
$$+ 0.67 \times 1.33 \times 9.83/27 + 0.67 \times 0.67 \times 9.83/27 \qquad = 10.8 \text{ CY}$$

As the concrete in each length of wall is taken off, the volumes of wall openings are deducted, and the dimensions of the openings are listed in a

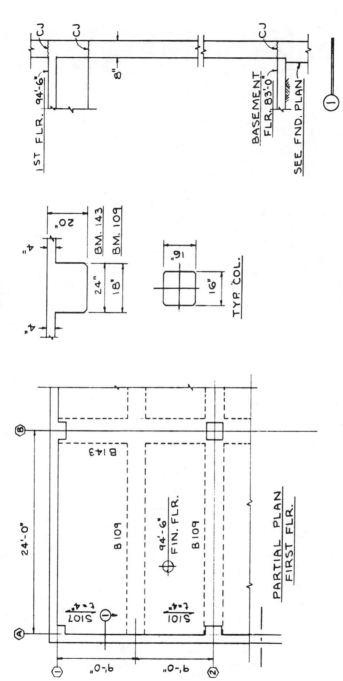

Figure 5.2 Beam and slab floor.

Figure 5E Wall forms, Scaffold jack staging.

single column on the take-off sheet. The information in this column is used to determine the cost of the 2-inch material and of the labor required to form or block out the openings.

In general, no deduction in wall form areas is made for small openings or for doorways or windows up to 8 feet in width. In such cases the gross form area must be set and stripped, and the wood bucks used to form the block-outs entail additional work. Where window openings occur, the plywood is removed from at least one face of the forms at the opening to enable concrete to be placed and vibrated below the sill.

Where wide entryways are encountered, the spandrel wall or beam above is supported on shoring, and only the actual contact area of the spandrel wall or beam is listed. In such cases the linear feet of spandrel wall or beam soffit are listed on the take-off, and constitute an additional cost item on the estimate sheet.

Where wide window openings occur, the wall forms are set full height, although there will be no plyform on the studs at the openings. No deduction in form area is made, and the spandrel wall above is supported by cross-pieces nailed to the wall form studs. The linear feet of such spandrel wall soffit is noted in the column for opening dimensions discussed above.

Where spandrel walls occur in connection with structural slabs, it may be possible to set a keyway for the slab and to form the spandrel wall full height. In such cases the structural slab is poured after the spandrel wall is stripped; as a general rule, savings are realized when job conditions permit this method of construction.

5.4.2   Unit Costs

Unit costs for concrete building walls are shown in Table 5.2.

The cost to set and strip staging, when required, is in addition to the costs shown for setting and stripping walls, and an estimated 1.1 MFBM of staging lumber will be required per MSF of wall face covered by the staging.

The unit costs shown for setting and stripping wall forms include the costs of handling scaffold jacks. The rental charges for the scaffold jacks are carried in the general costs section of the concrete estimate.

The unit cost shown for setting and stripping wall blockouts is based on the total amount of lumber required including the interior braces and struts. In practice, the estimator determines the cost to block out typical door, window, and duct openings, and the list of openings from the take-off is reduced to an estimated number of typical openings for pricing purposes.

Building walls must usually be poured before setting the soffit forms for the floor next above. When concrete buggies are used, the cost to set and move the elevated runways represents a substantial portion of the pouring cost, and demonstrates the advantages of using tower cranes or concrete pumps when tall buildings with large volumes of wall concrete are involved.

The cost to set and move buggy runways may be based on an estimated area of runways to be furnished, and this varies with the layout of the walls.

5.4.3   Form Lumber

On small buildings, it may be necessary to make wall form panels for all the walls, since little or no re-use may be possible. Forms must be made for a minimum of 2½ wall pours, and in most cases at least 5 MSF will be required.

On large buildings, the area of wall forms required may be approximated by dividing the total contact area of all the walls by an estimated 6 re-uses for the forms. Thus, materials will be furnished for, and the unit cost for making forms will be applied to, 1/6 of the total wall area. 1.17

Table 5.2. Unit Costs for Building Walls

Activity	Labor		Rental		Unit of Measure
	Unit Cost	Index	Unit Cost	Index	
Make form panels					
Walls (uniform)	$ 0.25	$K_{6+2} = 1000$	—		SF
Walls (irregular)	$ 0.33	$K_{6+2} = 1000$	—		SF
SS basement walls H10–H12	$ 1.65	$K_{6+2} = 1000$	—		SF
SS 1st through 4th floor walls H8–H14	$ 1.80	$K_{6+2} = 1000$	—		SF
SS 1st floor walls H14–H20	$ 2.00	$K_{6+2} = 1000$	—		SF
SS one face for gunite H9	$ 1.08	$K_{6+2} = 1000$	—		SF
SS wood or light steel staging to H40	$270	$K_{6+2} = 1000$	—		MSF[a]
SS planter walls H3	$ 1.71	$K_{6+2} = 1000$	—		SF
SS blockouts or bucks for openings	$1500	$K_{6+2} = 1000$	—		MFBM
SS spandrel beam soffit W16"	$ 6.00	$K_{6+2} = 1000$	—		LF
Pour building walls[b]	$ 8.00	$K_L = 13.48$	$10.50	$K_{Cr} = 75$	CY
Additional—set and move runways	$ 0.35	$K_{6+2} = 1000$	—		SF[c]
Pour planter walls[d]	$14	$K_{L+F} = 29.58$	—		CY

[a] Per MSF of wall face covered by the staging.

[b] Unit costs assume pour location is accessible to the crane, and that the concrete can be deposited directly into the forms. When pours are large or average in size, the unit cost for the crane includes 1 hour of travel time. On small pours, travel time will be additional. See section 5.1.4.3 for discussion of pour costs when concrete pump is used.

[c] Per SF of runway area.

[d] Costs to hoist or buggy concrete, when required, are additional.

Figure 5F Probation Department Building, Bakersfield, California. Contractor: Tumblin Company.

MSF of plyform and 2.0 MFBM of lumber will be required per MSF of wall form panels furnished.

An additional unit cost to cover the labor required to modify the forms for re-use must be applied to the remaining 5/6 of the total contact area of the walls. This unit cost may amount to 30 percent of the unit cost used to make the forms, and the percentage used will depend on the amount of variation in the successive wall pours and on the estimator's judgment.

5.5 Columns

Unit costs for columns are shown in Table 5.3.

Columns are poured from the floor soffit forms before installation of the floor reinforcing steel. When concrete buggies are used, no allowance for setting or moving runways is necessary.

Column pours entail small amounts of concrete, which is placed in three or more lifts to prevent excessive form pressures. Hoisting costs are reduced when columns are poured in conjunction with other larger pours, in which case the hoisting cost for the larger pour governs.

Excessive fluid pressure in sonotube forms will result in higher costs for additional finishing and for stripping. When high lifts are combined with

Table 5.3.   Unit Costs for Building Columns

Activity	Labor		Rental		Unit of Measure
	Unit Cost	Index	Unit Cost	Index	
Make form panels	$ 0.35	$K_{6+2} = 1000$	—		SF
SS columns (16×16) H8–H12	$ 2.10	$K_{6+2} = 1000$	—		SF
SS sonotube (16″ Φ) H8–H12	$ 4.00	$K_{6+2} = 1000$	—		SF
Pour with crane	$12	$K_L = 13.48$	$28		CY

Note: The unit costs for pouring assume normal access for the crane, and that the concrete can be deposited directly into the forms. When 11 CY or more of columns are poured at one time, the unit cost for the crane will include 1 hour of travel time. See Section 5.1.4.3 for discussion of pour costs when concrete pump is used.

maximum vibration, the unit cost to set and strip sonotubes may run as high as $8.00 per linear foot, based on $K_{6+2} = 1000$. When sonotubes are used, the pouring and stripping operations should always be properly supervised.

Since columns are formed in conjunction with the floor soffit forms, the amount of column forms required will depend on the area of floor forms furnished. No allowance for modifying column forms for re-use is necessary.

5.6   Beam and Slab Floors

A partial plan of a beam and slab floor is shown in Figure 5.2, and typical formwork is shown in Figure 5.3.

5.6.1   The Concrete Take-off

Structural plans designate concrete beams by width and depth of the beam cross-section. In all cases the width is listed first; the depth is always taken from the beam soffit to the top of the floor slab.

Beams 109 are 18″×20″ beams having 2.67 SF of beam sides per LF; beam 143 is a 24″×20″ beam having 2.67 SF of beam sides per LF.

The gross soffit area to which the unit costs are applied includes the areas of both the slab and the beam soffits. Thus, the gross soffit area enclosed by the building walls and by column lines B and 2 is 18×24 = 432 SF; no deductions are made for the plan areas of the pilasters or columns.

Figure 5G Column pour. (Photograph by The Burke Company).

We will designate that exterior wall formwork located between the upper two construction joints shown on section 1, Figure 5.2, as wall extension. This formwork must be set and stripped in conjunction with the formwork for the floor. The area of the wall extension forms shown on section 1 amounts to 1.67 SF per LF of wall. The formwork required for the interior face of the wall between the same two construction joints is included with the areas of the beam sides, and amounts to 1.33 SF per LF of wall.

The volumes of openings in the floor slab are deducted from the gross volume of concrete at the time the areas of the edge forms required for the openings are taken off.

Figure 5H Underground Garage, Beam and slab construction. Contractor: Tumblin Company.

The take-off sheet is set up in a similar manner to those shown in Chapter 4. There will be separate columns for at least the following items:

1. Miscellaneous.
2. SF of deck or deck opening edge forms.
3. LF of beams.
4. Number of form ties when required.
5. SF of wall extension.
6. SF of beam sides.

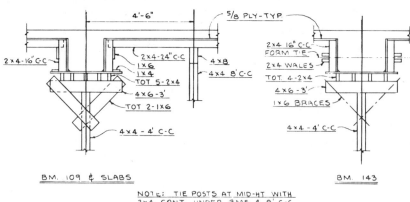

Figure 5.3 Forms for beam and slab floor.

7. Gross soffit area.
8. CY of concrete in the beams and slab combined.

   Items such as curbs, locker bases, and cement base are included in the floor concrete take-off when they occur. Unit costs for these items are discussed in Section 5.11.

5.6.2 Unit Costs

Unit costs for beam and slab floors are shown in Table 5.4.

The unit costs shown for setting and stripping beam sides and wall extensions include the costs of making the forms. The unit cost shown for setting and stripping the gross soffit includes the cost to set and strip all required falsework and soffit plyform.

The unit cost shown for pouring includes the cost of a steel trowel finish and assumes that a rotary finishing machine is used. When only a wood float finish is required the pouring cost will be reduced by about $1.50 per CY, based on $K_F = 16.10$.

5.6.3 Form Lumber

The amount of falsework and slab joists required is based on a preliminary design for a typical section of the floor. 1.5 MFBM of 2×4 is allowed per MSF of beam sides. About 1.17 MSF of plyform will be required per MSF of theoretical plyform area, and this area will be greater than the gross soffit area when the plyform for the beam soffits extends past the sides of the beams.

In general, about 2.6 MFBM of falsework lumber will be required per MSF of gross soffit area for H_c10, and the amount will increase about 4 percent per foot of increased height to H_c14. Similarly, the amount of plyform required will be about 1.3 to 1.5 MSF per MSF of gross soffit area.

In most cases, form materials are furnished for the largest floor, complete, and this material is re-used on the remaining floors. Depending on the areas and number of the succeeding floors, on the soffit finish required, and on the variations in beam sizes and spacings, the estimator may allow additional plyform for some of the additional floors.

Re-shoring is usually required for one or more floors immediately below the floor being formed. The amount of re-shoring will depend on the load from the floor being poured and on the design live loads for the floors below. The live load assigned to the floor being poured will depend on the estimator's judgment, and 20 PSF has been used.

Table 5.4. Unit Costs for Concrete Floors

Activity	Labor		Unit of Measure	Production Rate (CY/hr)
	Unit Cost	Index		
SS wall extension	$ 3.10	$K_{6+2} = 1000$	SF	—
SS deck edge and blockouts	$ 2.30	$K_{6+2} = 1000$	SF	—
SS slab pour bulkheads	$ 2.60	$K_{6+2} = 1000$	SF	—
Beam and slab floors				
SS beam sides	$ 2.10	$K_{6+2} = 1000$	SF	—
SS gross soffit forms H_c10[a]	$ 1.65	$K_{6+2} = 1000$	SF	—
Pour and trowel finish[b]	$16.00	$K_{L+F} = 29.58$	CY	12
Concrete joists with steelforms				
SS falsework[c]				
Pour and trowel finish	$15.00	$K_{L+F} = 29.58$	CY	15
Structural slabs				
SS soffit forms H_c10[d]	$ 1.65	$K_{6+2} = 1000$	SF	—
Pour and trowel finish[b]	$ 0.40	$K_{L+F} = 29.58$	SF	10
plus	$ 5.00	$K_L = 13.48$	CY[e]	

[a] Gross soffit area = area of beam soffits plus area of slab soffits = total area of beam and slab floor. Increase unit cost 4.5 percent per foot of additional height to H_c14.
[b] The unit costs for pouring concrete assume the concrete is poured directly from transit-mix trucks, or is deposited directly into the forms with a crane or a concrete pump. The costs to hoist the concrete, or to buggy or pump it horizontally, are additional. Included are the costs to set screeds, pour complete, cure, clean up, and cover the slab with building paper on completion.
[c] Use the bridge falsework curve, Figure 4.16, and increase these costs by 7 percent.
[d] Increase unit costs 4.5 percent per foot of increased height to H_c14.
[e] Per CY in excess of 12.3 CY per MSF of slab.

238

Typical re-shoring may consist of 4×4 posts at midspan for each beam, and under the girders at the points where the beams frame into them. Re-shoring is set after the floor has been stripped, and the material required is normally in addition to the falsework required for the floor forms. Wedges must be furnished for each post, and the number of wedges is based on the MFBM of posts required.

5.7 Concrete Joist Floors

Concrete joist floor systems constructed with steel forms on wood falsework are frequently encountered. One-way concrete joist floors are similar to beam and slab construction in appearance. Two-way concrete joist floors are built with steel "dome" forms, and the concrete joists run in both directions on 2- to 5-foot centers. Both types of floors are, in effect, deep slabs with voids in the bottom face that are created by the steel forms. Such floors are often referred to as pan slabs or dome slabs.

Typical steel forms for concrete joist construction are shown in Figure 5.4, and a typical section through a one-way concrete joist floor is shown in Figure 5.5, together with the required wood forms and falsework. The volumes of the voids created by various sizes of typical steel forms are given in Table 5.5.

The steel forms shown in Figures 5.4 and 5.5 are adjustable. The pans pass by, and are nailed to, the sides of the joist soffit forms to attain the desired joist depth. Fixed depth pans are sometimes used. These have out-standing flanges at the bottom edge that rest on, and are nailed to, the joist soffit material.

The steel forms or pans are normally furnished, set, and stripped by subcontractors at a unit price per square foot of gross slab area. These subcontractors usually quote each job two ways. One quotation includes the costs of furnishing, setting, and stripping the pans on the general contractor's falsework. The second quotation includes the cost of doing this work on falsework that is furnished, set, and stripped by the subcontractor.

5.7.1 The Concrete Take-off

On a typical concrete joist floor system, the bulk of the pans used are of uniform width. To fit the voids to the plan dimensions of the slab, however, it is usually necessary to use one or two rows of narrower pans at each side, and sometimes some of the joist widths are varied for the same reason. The volume of concrete in the floor is determined by deducting

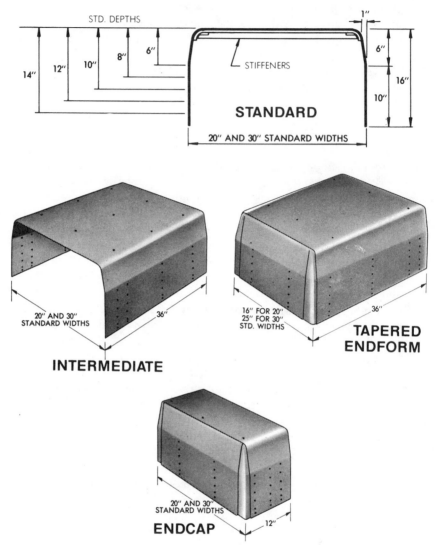

Figure 5.4 Details for Ceco Adjustable Steelforms, (The Ceco Corporation).

the volumes of the voids created by the pans from the volume of the solid, full-depth slab.

Wall extensions and some beam sides must always be formed in connection with pan slab work. In Figure 5.5, 2.00 SF of wall extension are required per linear foot of wall. In addition, 0.54 SF of beam sides per

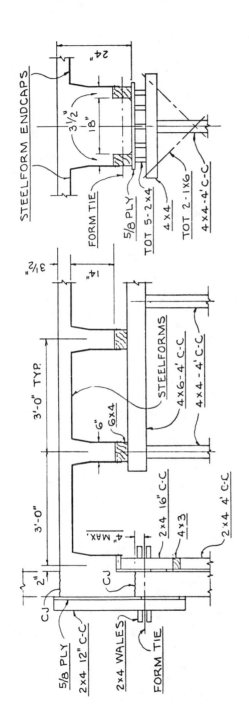

Figure 5.5 Concrete joist floor.

241

Table 5.5.   Voids Created by Various Size Ceco Adjustable Steelforms[a]

Depth of Steelform	Cubic Feet of Voids per Linear Foot for Various Widths of Adjustableforms				
	40″	30″	20″	15″	10″
		Standard			
6″		1.207	0.790	0.582	0.373
8″		1.623	1.068	0.790	0.512
10″		2.040	1.345	0.998	0.651
12″		2.457	1.623	1.207	0.790
14″		2.873	1.901	1.415	0.929
		Deepset[b]			
12″	3.227	2.394	1.561	1.144	0.727
14″	3.783	2.811	1.838	1.352	0.866
16″	4.338	3.227	2.116	1.561	1.005
18″	4.894	3.644	2.394	1.769	1.144
20″	5.450	4.061	2.672	1.997	1.283
22″	6.005	4.477	2.950	2.186	1.422
24″	6.561	4.894	3.227	2.394	1.561

[a]These data apply to Ceco Adjustable Steelforms for one-way joist floor construction, and are furnished through the courtesy of The Ceco Corporation.
[b]The tapered endform for 40″ wide deepest adjustableforms is 3 feet long and tapers to 34 inches in width. See also Figure 5.4.

linear foot of wall are required for the interior face of the wall, and 1.08 SF of beam sides are required per linear foot of the 24-inch-deep girder.

5.7.2   Unit Costs

The unit costs for concrete joist floors are shown in Table 5.4. The unit costs for setting and stripping wall extensions, slab edge forms, and pour bulkheads are the same for beam and slab construction and for concrete joist floor systems. The unit cost to set and strip beam side forms is higher in the case of concrete joist floors because small areas are usually involved. For situations like those shown in Figure 5.5, the unit cost to set and strip the beam sides is about 1.5 times the unit costs shown for beam sides in connection with beam and slab floors.

The unit cost to pour concrete joist floors is less than the unit cost to pour beam and slab floors because the slab and joists are normally poured in one pass. When a wood float finish is required, the pour cost may be reduced by $0.03 per SF of gross slab area, based on $K_F = 16.10$.

5.7.3 Form Lumber

The amount of falsework required is based on a preliminary design for a typical section of floor, and includes the posts and beam system necessary to support the beam forms at the interior faces of the walls.

In general, about 7.0 MFBM of falsework lumber will be required per MLF of joists and beams for one-way concrete joist construction, when $H_c = 10$. This quantity should be increased 4 percent per foot of increased height to H_c14. Two-way concrete joist systems require about 2.3 MFBM of falsework per MSF of gross slab area when $H_c = 10$, and this quantity should also be increased 4 percent per foot of increased height to H_c14.

5.8 Structural Slabs

Structural slabs vary greatly in area and depth, and the amount of falsework required is based on a preliminary design for a typical section in all cases.

In general, about 2.2 MFBM of lumber will be required per MSF of soffit area for H_c10, and this quantity increases about 4 percent per foot of increased height to H_c14. 1.17 MSF of plyform will be required per MSF of soffit area.

Unit costs for structural slabs are shown in Table 5.4, and are based on slab areas of 800 SF or greater, per slab. The unit costs to set and strip soffit forms for small slabs are higher and may amount to as much as $3.00 per SF for slabs 50 to 100 square feet in area, based on $K_{6+2} = 1000$.

5.9 Stairs

A section of a typical concrete stair, and the required formwork, is shown in Figure 5.6. Stairwell walls are treated as typical building walls, and are not included in the stair estimate, since the stairs are constructed after these walls and the floors on them have been completed.

Material from wall and floor formwork will be available for re-use in many cases, and no additional allowance for stair soffit form materials may be necessary. In general, about 3.2 MFBM of soffit form lumber will be required per MSF of stair, or stair plus landings, plan area.

In all cases, 2-inch material must be furnished for the riser forms, and

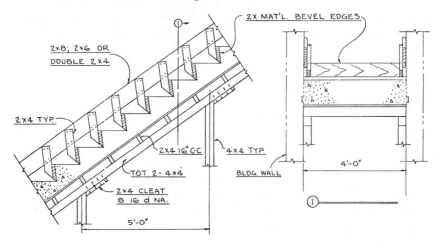

Figure 5.6 Stair forms.

these forms should be firmly secured, since they must support the workmen during the pour. In addition, the riser forms are subjected to horizontal forces resulting from the fluid pressure of the concrete.

The support details for the riser stringers will vary to suit the existing conditions. The stringers are supported with removable screed-type supports, set square with the soffit, on 6- to 8-foot centers. The stringers must be restrained horizontally, and vertically to prevent upward rotation. Temporary struts to previously poured concrete are often used for these purposes.

When stairs are poured in a single operation, as opposed to being poured rough and topped later, the riser forms are removed during finishing, and the cost of this work is included in the pour costs.

5.9.1   Unit Costs

Unit costs for stairs are shown in Table 5.6.

The unit cost for setting and stripping stairs with no landings assumes single flights formed between previously poured slabs or beams. The unit cost for placing stair topping includes the cost of setting and stripping temporary riser forms, and is applied to the total area topped, including the area of the risers.

5.10   Tilt-up Walls

In this type of construction, the walls are cast in a horizontal position at the site, and are tilted up and moved onto previously poured foundations

Table 5.6. Unit Costs for Concrete Stairs

| Activity | Labor | | Unit of Measure |
	Unit Cost	Index	
SS soffit forms			
Stairs with no landings	$ 3.50	$K_{6+2} = 1000$	SF[a]
Stairs plus landings	$ 3.95	$K_{6+2} = 1000$	SF[a]
SS riser forms	$ 2.35	$K_{6+2} = 1000$	LF
SS Hl stair edge	$ 3.45	$K_{6+2} = 1000$	SF
Pour and finish complete[b]	$90	$K_{L+F} = 29.58$	CY
Pour rough[b]	$30	$K_{L+F} = 29.58$	CY
Place and finish topping[b]	$ 3.75	$K_{L+F} = 29.58$	SF
Place and finish concrete in metal pans[b]	$ 1.20	$K_{L+F} = 29.58$	LF
Set 3″ metal tread nosing	$ 1.60	$K_{L+F} = 29.58$	LF
Additional for tread abrasives	$ 0.15	$K_{L+F} = 29.58$	SF
Set 3″ sleeves for rail posts	$ 1.85	$K_{6+2} = 1000$	EA
Set pipe stair railing			
Grouted posts	$27	$K_{6+2} = 1000$	EA
Brackets (2 bolts ea.)	$21	$K_{6+2} = 1000$	EA

[a]Stair soffit area is based on slope lengths.
[b]Costs to hoist and to buggy concrete are additional. Estimated production rate is 2 CY per hour.

with a truck crane. After the panels have been erected and braced, columns or pilasters are cast in place between them. In the following discussion of tilt-up panels, the terms "vertical," "horizontal," and "above" will refer to positions on the erected panels.

5.10.1 General Costs

The following general cost items must be considered in addition to those presented in the Chapter 4 checklist. The rental and material costs for these items are carried in the general costs section of the concrete estimate, but the installation costs, if any, are carried in the operating sections of the estimate.

5.10.1.1 CASTING SITE. It may be necessary to acquire additional space for a casting site. Maximum economy is achieved when the panels are cast on exterior grade slabs which will become part of the finished construction. In such cases, the panels are cast with their bottom edges about 5 feet from, and parallel to, their erected location. The truck crane is

Figure 5I Tilt-up building. Panels 5½ inches thick by 38 feet in height. Contractor: Tumblin Company.

located inside the building, and each panel can be tilted up and erected without moving the crane during the process.

Since the wall foundations must remain exposed until the panels are erected, a construction joint in the exterior slab is required; it will be parallel to, and about 3 feet from, the wall line.

When exterior slabs are not available, wall panels are cast on the interior grade slab if space permits, and the crane is located outside the building. As before, a construction joint will be required in the floor slab, near the wall line.

Space is usually limited, and panels are frequently cast in stacks of two or three panels. The stacks should be low enough to permit the concrete to be placed directly from the transit mix truck. When panels are stacked, the locations of dowels and wall openings, as well as the plan dimensions of the panels themselves, must be compatible with this procedure.

As a last resort, wall panels may be cast on temporary, 4-inch-thick waste slabs. The costs to furnish and remove the waste slabs are included in the operating section of the estimate.

Casting bed slabs must be smooth, uniform, plane surfaces. Small imperfections, permissible in grade slabs, may be objectionable when they appear on erected walls. Wall panels should not be cast over cracks, construction joints, or temporarily filled expansion joints, since this will usually result in cracked wall panels.

5.10.1.2 ENGINEERING.   The wall panel thickness and the reinforcing steel shown on the structural plans will have been designed to enable the panel to function as a portion of a building wall. The procedures required to cope with the additional stresses encountered during tilting are the responsibility of the general contractor.

The necessary engineering services are provided by the suppliers of the concrete accessories used in tilt-up construction. These firms submit quotations covering the costs of their engineering services and the material and rental costs for the accessories required.

During the erection process, the wall panels are tilted upward to a nearly vertical position about an axis of rotation located at their bottom edges. They are then lifted and placed in their final position on the building foundations.

As a panel is tilted, it is supported by the casting slab at its bottom edge and by crane rigging attached to from 2 to 16 lift-points located on the face of the panel. The lifting shackles, or lift hardware, are bolted or pinned to inserts which have been cast in place in carefully designed locations.

The lift-point inserts are located to keep the bending stresses, which occur during erection, within allowable limits. The inserts are located in horizontal rows of 2 or 4 inserts each, and from 1 to 4 rows will be required. When but one row is required, 2, 3 or 4 inserts may be used, and these are sometimes located in the top edges of short panels.

A panel requiring 4 rows of 2 inserts each is designated as a 2-wide by 4-high pick-up or as a 4-row, 8-point pick-up. A 2-wide by 4-high pick-up is illustrated in Figure 5.7. The tabulated lift-point dimensions will be varied slightly by different designers.

For design purposes, the vertical elements of the panel are assumed to be continuous beams supported by the ground reaction at the bottom and by assumed transverse beams which are co-incident with the horizontal rows of inserts. These horizontal rows are spaced vertically to equalize the bending moments as much as possible.

The transverse beams are supported at the lift-points, and the inserts in the horizontal rows are spaced to equalize the bending moments in the

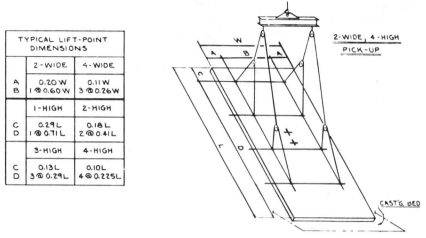

TYPICAL LIFT-POINT DIMENSIONS		
	2-WIDE	4-WIDE
A	0.20 W	0.11 W
B	1 @ 0.60 W	3 @ 0.26 W
	1-HIGH	2-HIGH
C	0.29 L	0.18 L
D	1 @ 0.71 L	2 @ 0.41 L
	3-HIGH	4-HIGH
C	0.13 L	0.10 L
D	3 @ 0.29 L	4 @ 0.225 L

Figure 5.7 Rigging for tilt-up panels.

transverse beams. A mat of number 4 bars is frequently required in the upper surface of the slab at each horizontal row to increase the effective width of the transverse beam and the resistance of the inserts in tension and in shear.

The center of gravity of the lift-insert load pattern must be located on a vertical line through the center of gravity of the wall panel, and should be at least 18 to 24 inches above the center of gravity of the panel. In determining the latter, the areas of all wall openings must be considered. A single rigging system should be used for all panels, when possible. The cost of any additional inserts that may be required will be small compared to the cost of re-rigging the crane during erection.

Typical wall panels are 5½ to 6½ inches thick, and are too thin to function as reinforced concrete during the erection process. Paradoxically, the panels are considered as un-reinforced plain concrete, the erection stresses are calculated by the flexure formula, and the working stress used is a percentage of the modulus of rupture for the concrete.

In all cases, test cylinders should be taken from the wall panel concrete and, when facilities are available, test beams should be cast to determine the modulus of rupture directly.

Crushed aggregates normally produce stronger concrete and, when used, the ratio of the modulus of rupture to the compressive strength is generally greater.

The following rules of thumb should be considered in tilt-up panel designs.

1. Lift-point inserts should never be spaced more than 13 feet apart, nor less than 18 inches from the edges of openings or of the panels.
2. The length of the cantilever overhangs at the top and sides of panels should not exceed (t-1) feet, where t = the thickness of the panel in inches.
3. The number of lift-point inserts used on a given panel should not be less than the total panel weight divided by the lesser of the safe working loads for the inserts in tension or in shear.

5.10.1.3 HIGH STRENGTH CONCRETE. Concrete of a higher strength than that specified may be required to attain the necessary modulus of rupture, or to attain it at an earlier age.

5.10.1.4 CURE AND BOND BREAKER. The same compound must be used to cure the casting bed concrete and to break the bond between it and the wall panel. This material is furnished by the concrete accessories supplier, and must be carefully applied as directed. Two coats are required, and the material is applied with sprays at a rate of about 500 square feet per gallon, per coat. The second coat is applied the day before the reinforcing steel is placed, and the surface must be checked and touched up, if necessary, after the steel has been installed. Polyethylene sheets or paper are generally unsatisfactory bond-breakers; wrinkling will occur in most cases.

5.10.1.5 ADDITIONAL REINFORCING STEEL. In addition to the mats that may be required at the transverse beams, extra reinforcing steel may be necessary at the sides, top, or bottom of large or unsymmetrically placed wall openings, to resist erection stresses. This additional reinforcing is not included in the quotations received from reinforcing steel subcontractors, and the costs to furnish and install it are included in the general costs section.

5.10.1.6 STRONGBACKS. Strongbacks are often required when large openings occur in the wall panels, and may be located in horizontal or vertical positions. They ordinarily consist of pairs of timber beams or steel channels, tightly bolted to inserts cast into the wall panels. Strongbacks are designed to take all the bending moments in a given area, and the bending resistance of the wall panel concrete in that area is neglected. After the panels have been erected, the strongbacks may be removed for re-use.

5.10.1.7 SHIMS. To ensure that the bottom edges of the erected panels are set level, at least two sets of steel or masonite shims are required under each panel.

5.10.1.8 TEMPORARY BRACES. At least two braces are required for each of the wall panels, to plumb and to support them after erection. The bracing system is designed to withstand the anticipated wind loads, and remains in place until the floor or roof diaphragms have been installed. Long braces require strut and tie supports at their midpoints.

Telescoping tubular steel braces having screw jacks for easy adjustment are furnished by the concrete accessories suppliers on a monthly rental basis. These braces are available in lengths ranging from about 16 to 40 feet. Steel end-hardware with screw jacks attached is also available for use with double 2×4's or 2×6's furnished and fabricated by the contractor. When panels as high as 40 feet must be erected, a bracing system of wire rope, turnbuckles, and dead-men may be used in addition to the rigid braces.

For maximum economy, it is essential that the panel casting and erection sequence selected permit the upper ends of all braces to be bolted to the horizontal panels prior to their erection, and that the grade slab inserts be clear and unobstructed when needed.

The labor cost to connect the braces to the horizontal panels and the cost to remove the braces on completion are based on a single unit cost for both operations. The labor cost to bolt the braces to the grade slabs and to install struts when required are included in the costs to erect the panels.

5.10.1.9 INSERTS AND BOLTS. This item covers the costs to furnish and install the inserts required for the lift hardware, the strongbacks, and the temporary braces.

5.10.1.10 LIFT HARDWARE. This item covers the rental charges for the lift plates and eyes that will connect to the lift-point inserts. Twist-lock lift hardware is available and, when used, is required for only one panel. When bolted lift hardware is used, it is necessary to furnish hardware for two or three panels to prevent delays during erection. The costs to install and to remove lift hardware are included in the erection costs.

5.10.1.11 FILL AND PATCH INSERT HOLES. The holes at all inserts must be filled and patched on completion, and the costs for this work are carried in the additional finishing item.

5.10.1.12 DRY PACK AT FOUNDATION. This item covers the costs to furnish and place dry pack material between the bottom edges of the wall panels and the wall foundations. The costs of materials and the labor are carried in the general costs section of the concrete estimate.

5.10.2 Form Lumber

In most cases, the only new material required is the 2-inch-thick lumber used for the edge forms. The bracing necessary to align and to support these forms is normally available at the job site.

5.10.3 Unit Costs

Unit costs for tilt-up wall panels are shown in Table 5.7.

The unit cost for pouring the waste slabs and the wall panels assume that concrete is delivered directly to the slabs in transit-mix trucks.

The unit cost and the production rate shown for pouring the columns represent an order of magnitude for this work, but good judgment must be used in selecting the unit costs to be used. Columns for tilt-up work are often small in cross-section, and contain a large amount of reinforcing steel; in addition to the typical column bars and ties, the horizontal wall panel reinforcing projects into the columns, and these bars may have hooked ends.

5.10.4 Erecting Tilt-up Panels

Wall panels should be erected by a rigging company experienced in this type of work. Such firms are normally paid on the basis of hourly rental rates for the truck crane and for the riggers. The crane rental rate includes all required spreader beams and rigging, and the crane used should have twice the theoretical lifting capacity required by the actual job conditions. As shown in Fig. 5.7, continuous wire rope rigging is used to equalize the lift-point loadings, insofar as possible, during erection.

Panels erected with the crane positioned inside the building should be cast with their inside faces up; panels cast on the building floor slab, and erected from the outside, should be cast with their outside faces up. This procedure obviates the dead man's pick, which occurs when a panel would fall towards the crane if the rigging were to fail during the erection process.

Two to four extension ladders are needed to enable the workmen to disconnect the rigging from the erected and braced panels.

Table 5.7.   Unit Costs for Tilt-Up Panels

| Activity | Labor | | Unit of Measure |
	Unit Cost	Index	
Waste slabs			
FG	$ 0.06	K_L = 13.48	SF
SS edge forms H4″	$ 0.75	K_{6+2} = 1000	LF
Pour and trowel finish[a]	$ 0.32	K_{L+F} = 29.58	SF
Wall Panels			
SS edge forms (solid)	$ 1.15	K_{6+2} = 1000	LF
SS edge forms (2-piece)	$ 1.40	K_{6+2} = 1000	LF
SS blockout forms	$ 1.75	K_{6+2} = 1000	LF
Set inserts	$ 5	K_{6+2} = 1000	EA
Pour and trowel or broom finish[a]	$ 0.40	K_{L+F} = 29.58	SF
plus	$ 5	K_{L+F} = 29.58	CY[b]
Columns (poured between panels)			
Make forms	$ 0.40	K_{6+2} = 1000	SF
SS column forms	$ 3.10	K_{6+2} = 1000	SF
Pour columns[c]	$20	K_L = 13.48	CY
Set shims	$ 1.90	K_{6+2} = 1000	EA
SS temporary braces	$11	K_{6+2} = 1000	EA
Dry-pack panels at bottom	$35	K_L = 13.48	CF

[a]These unit costs assume the concrete is poured directly from transit-mix trucks.
[b]Per CY in excess of 18.5 CY per MSF of wall panel area.
[c]The cost to hoist the concrete is additional. Estimated production rate is about 2.5 CY per hour.

In rare instances, the bond breaker may not perform perfectly. In such cases a slight horizontal movement induced by a good hydraulic jack may be required.

Under optimum conditions, solid rectangular 2-wide, 2-high pick-ups may be erected at a rate of 3 panels per hour, and solid 4-wide, 4-high pick ups may be erected at a rate of 2 panels per hour. Large panels, irregular in shape, requiring strongbacks and additional care in handling, may require 1 hour for erection.

A minimum erection crew for 2 to 4 point pick-ups may consist of a rigger foreman, two riggers, and two carpenters. When 8 to 16 point pick-ups are involved, two additional riggers should be used. When strongbacks must be removed for re-use, two additional carpenters and, in some cases, an A-frame truck are required for only that time necessary to do this work.

Figure 5J Tilt-up panel erection, (Photograph by The Burke Company).

5.11   Grade Slabs and Miscellaneous Items

Typical details are shown in Figures 5.8 and 5.9. Unit costs for this work are given in Tables 5.8 and 5.9.

The cost to pour concrete slabs is based on two separate and distinct cost factors—the cubic yard cost to place the concrete, and the square foot cost to finish the surface of the slab. The unit cost per square foot shown in Table 5.8 includes the costs to place and finish the concrete in a 4-inch-thick slab, which contains 12.35 CY per MSF.

The square foot cost for grade slabs assumes the use of a rotary finishing machine, and includes the costs of setting screeds, curing the slab, and covering it with building paper on completion. The square foot price for sidewalks assumes hand finishing, and includes the costs of a light trowel or broom finish, as well as the costs of joining, edging, and curing.

The cubic yard cost is in addition to the square foot cost, and includes the costs to place that additional concrete contained in scoop footings, thickened edges, and stud wall curbs or in slabs greater than 4 inches in thickness. The unit cost shown for forming, pouring, and finishing stud wall curbs represents an additional cost for this work, over and above the cubic yard cost discussed above.

All grade slab concrete is carried in a single column on the take-off sheet, and there will be separate columns for the slab areas, edge forms,

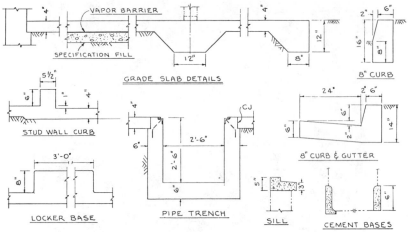

Figure 5.8 Grade slabs and miscellaneous items.

stud wall curbs, and hand excavation. Specification fill, vapor barrier, and mesh are carried in the miscellaneous column. The volume of concrete in excess of 12.3 CY per MSF of slab area is noted below the total for the column containing the slab areas.

When specification fill is required, the material is purchased by the ton, and its compacted density will range from 1.5 to 1.7 tons per cubic yard. Two fine grades must be made—one for the fill, and a second one for the slab.

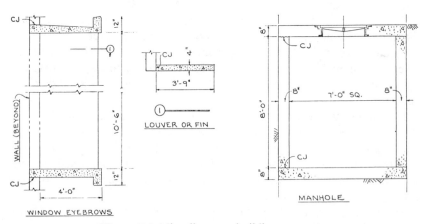

Figure 5.9 Miscellaneous building concrete.

Table 5.8. Unit Costs for Grade Slabs and Curbs

Activity	Labor		Rental		Unit of Measure
	Unit Cost	Index	Unit Cost	Index	
FG for specification fill	$ 0.06	$K_L = 13.48$	—		SF
FG for concrete slabs	$ 0.08	$K_L = 13.48$	—		SF
Place specification fill	$ 6.50	$K_L = 13.48$	$4	$K_{BF} = 31$[a]	CY
SS edge forms for slabs or walks	$ 2.25	$K_{6+2} = 1000$	—		SF
Set 6×6/10×10 wire mesh	$75	$K_{6+2} = 1000$	—		MSF
Place vapor barrier	$45	$K_{6+2} = 1000$	—		MSF
Hand excavation at scoops	$14	$K_L = 13.48$	—		CY
Pour and finish grade slabs[b]	$ 0.33	$K_{L+F} = 29.58$	—		SF
plus	$ 4.10	$K_L = 13.48$	—		CY[c]
Set and move buggy runways	$ 0.06	$K_{6+2} = 1000$	—		SF[d]
Pour and finish sidewalks	$ 0.36	$K_{L+F} = 29.58$	—		SF
plus	$ 4.50	$K_L = 13.48$	—		CY[c]
SS, P & F stud wall curbs	$ 4.00	$K_{L+F} = 29.58$	—		LF
SS, P & F 8" curb and gutter	$ 4.80	$K_{L+F} = 29.58$	—		LF
SS, P & F 8" curb (straight)	$ 2.70	$K_{L+F} = 29.58$	—		LF
SS, P & F 8" curb (irregular)[f]	$ 5.00	$K_{L+F} = 29.58$	—		LF

[a] K_{BF} = the sum of the hourly rental rates for a ½ CY wheel loader and a hand-operated vibratory compactor.
[b] These unit costs assume the concrete is placed directly from transit-mix trucks. The cost to buggy or pump concrete horizontally is additional. The unit costs include the costs for setting screeds, pouring complete with a trowel finish, curing, cleaning up, and covering the slab with building paper.
[c] Per CY in excess of 12.3 CY per MSF of slab.
[d] Per SF of slab area.
[e] Similar to b, above, except that walk is not covered with paper.
[f] Such as curbs for parking lot islands.
Note: Curb and gutter details are shown in Figure 5.8.

255

Table 5.9. Unit Costs for Miscellaneous Items

Activity	Labor Unit Cost	Index		Unit of Measure
Pipe trenches				
Fine grade	$ 0.23	K_L	= 13.48	SF
Make, set, and strip inside face	$ 3.30	K_{6+2}	= 1000	SF
Pour and finish, including slab	$32	K_{L+F}	= 29.58	CY
Window sills				
Set and strip	$ 4	K_{6+2}	= 1000	LF
Pour and finish	$ 4	K_{L+F}	= 29.58	LF
Locker base				
Set and strip	$ 2.45	K_{6+2}	= 1000	SF
Pour and finish	$19	K_{L+F}	= 29.58	CY
Cement base				
6″ Straight, including wall chase	$ 2.50	K_F	= 16.10	LF
6″ Cove, including wall and floor chases	$ 6.00	K_F	= 16.10	LF
Manholes[a] FG, SS and P complete	$150	K_{C+L+F}	= 45.92	CY
Fine grade	$ 0.28	K_L	= 13.48	SF
Set and strip bottom slab	$ 2.30	K_{6+2}	= 1000	SF
Make wall forms	$ 0.25	K_{6+2}	= 1000	SF
Set and strip walls H9–H12	$ 2.00	K_{6+2}	= 1000	SF
Set and strip soffit form	$ 3.00	K_{6+2}	= 1000	SF
Pour and finish slabs	$25	K_{L+F}	= 29.58	CY
Pour walls	$16	K_{L+F}	= 29.58	CY
Deep pits[b]				
SS wall forms H18–H24	$ 1.30	K_{6+2}	= 1000	SF
SS wall forms H30	$ 1.55	K_{6+2}	= 1000	SF
Pour 12″ thick walls	$12	K_L	= 13.48	CY
Pour 8″ thick walls	$22	K_L	= 13.48	CY
Eyebrows				
SS soffit forms H_c4[c]	$ 1.85	K_{6+2}	= 1000	SF
SS soffit forms H_c10[c]	$ 2.70	K_{6+2}	= 1000	SF
Pour and finish slabs	$31	K_{L+F}	= 29.58	CY
Make forms for fins	$ 0.35	K_{6+2}	= 1000	SF
SS fins	$ 1.65	K_{6+2}	= 1000	SF
Pour fins[d]	$17	K_L	= 13.48	CY

[a] Cost to set MH ring and cover, cost to fit wall forms at pipes or ducts, and cost to set rungs are additional. K_{C+L+F} = hourly wage rate for 1 carpenter plus 1 laborer plus 1 finisher.
[b] Cost to make wall forms is additional.
[c] Cost to set and strip edges or beams is additional.
[d] Cost to hoist and/or buggy concrete is additional.

The unit costs shown for fine grading assume essentially flat or uniformly sloping slabs and sandy loam soil. When many grade breaks occur, these unit costs may increase by as much as 60 percent. All fine grading costs assume the machine grade to be within ±0.1 foot of finish grade.

The unit costs shown for sidewalks and curbs are typical for cases where the general contractor does the work with his own forces. When appreciable amounts of such exterior concrete are involved, savings can usually be achieved by having this work done by subcontractors.

6

Carpentry

There are two broad categories of carpentry work—rough carpentry and finish carpentry. For estimating purposes, these two types of work are taken off separately, but are priced in a single section of the job estimate.

The format for the carpentry estimate sheet is identical to the one described at the beginning of Chapter 4 for structure concrete, and the procedures followed to estimate the time required for completion are identical to those illustrated in the sample estimates in Chapter 4.

Rough and finish carpentry are usually covered in two separate sections of the specifications. In addition to the work encompassed by these sections, the installation costs for certain materials covered under other sections of the specifications are included in the carpentry estimate.

Thus hollow metal doors and frames, millwork, and finish hardware are all furnished, F.O.B. the job, under separate sections of the specifications. The costs of these materials are entered in the appropriate sections of the estimate summary sheet, but the costs for their installation are carried in the finish carpentry estimate. The suppliers of these materials make their own take-offs in all cases, tender their quotations on bid day, and are, in this respect, similar to subcontractors.

The lump sum quotation for millwork normally includes the costs of all wood doors, unfinished wood cabinets, milled trim pieces, and that rough lumber which requires milling and which is detailed in the job plans. In addition, the millwork quotation may include a square foot cost for hardwood finish-plywood.

Pre-finished wood cabinets, when specified, are covered under a separate section of the specifications. They are furnished and installed, complete, by the manufacturers on a sub-contract basis. Pre-finished wood cabinets are furnished with all doors installed, and no additional finish hardware, painting, or work by the general contractor is required.

The lump sum quotation for finish hardware will include the costs of all hardware required for metal doors, wood doors, and unfinished wood cabinets, and all accessories such as door closers, door stops, and metal thresholds.

Toilet stalls are covered in a separate section of the specifications, and are usually furnished and installed by specialty subcontractors. When quoted F.O.B. job, toilet stalls are installed under finish carpentry. In all cases toilet stalls are taken off by the suppliers.

6.1   General Costs

The general costs section for carpentry work will include the following items:

1. Lumber.
2. Plywood.
3. Miscellaneous iron.
4. Other materials.
5. Nails.
6. Material suppliers or subcontractors.
7. Hand power tools.
8. Hand tools.
9. Saw filing.
10. Compressor.
11. Develop power supply.
12. Table saw.
13. Hoisting.
14. Other items.

The rental periods used for equipment items on the above list cover only the time the equipment is required for carpentry work. Care must be exercised to avoid overlapping the rental periods for those items previously included in the general costs section for structure concrete, which may be available, as needed, for the carpentry work.

6.1.1.   Lumber and Plywood

Lumber is, in general, taken off piece by piece, and a combined lumber list is prepared and distributed to the lumber yards for materials quotations. In all cases the lengths are listed in multiples of 2 feet. Thus, 500 2×6 studs 8'-6" long would be listed as 2×6, 500/10. When these studs are spaced 16 inches on center, the waste will furnish one piece of fire blocking, and no additional allowance for waste would be necessary in connection with these two items.

Where materials such as ledgers or stud wall plates are taken off by the linear foot, an allowance for waste must be made, and its size will depend on the plan dimensions and on the estimator's judgment. Thus, if the average length of run were slightly greater than 30 feet, the estimator would plan to use two pieces 16 feet long, and the factor for waste would be approximately 32/30 = 1.07. In such a case, the neat linear feet taken off would be increased by 7 percent, and the resulting figure converted into an even number of pieces 16 feet long.

Where boards are used as sheathing or as decking, the finished board widths must also be considered when determining waste allowances.

For example, assume that a long section of stud wall, having an overall height of 9'-0", is to be covered with 1×6 diagonal sheathing. The sheathing is to be laid at 45 degrees, and the finished board widths are 5½ inches. The diagonal length will be 1.414 × 9.00 = 12.73 feet, and the required board length will be 12.73 feet plus 5½ inches = 13.19 feet.

The boards will be ordered in 14-foot lengths, and the waste factor applicable to the gross sheathing area will be, in this case, 14/13.19 × 6/5.5 = 1.16. A waste allowance of 20 percent is often used for diagonal sheathing.

Whenever possible, the plywood take-off is based on the actual number of sheets required. Thus, a roof area measuring 31 feet by 102 feet is considered as 32 × 104 = 3.328 MSF, and would require 104 sheets 4 feet × 8 feet in size.

On all jobs, there will be some sections where the allowances for waste cannot readily be determined. In such cases allowances of 10 percent for waste lumber (due to end cutting), and of 15 to 20 percent for waste plywood have been used.

6.1.2   Miscellaneous Iron

This item covers the purchase price, F.O.B. job, of the bolts, joist hangers, and other miscellaneous iron required for the carpentry section. The cost to set those bolts which are embedded in concrete is included in the operating section of the carpentry estimate relating to the lumber requiring the bolts. The costs to install bolts connecting wood to wood are included in the unit costs for the various types of carpentry work.

6.1.3 Nails

Nailing schedules for the major types of work are given in the plans and specifications. The theoretical number of nails required for a given type of work is estimated by considering a typical piece or section and extrapolating this figure, based on the ratio of the total board feet involved to the board feet in the piece or section considered. The theoretical amount of nails so determined should be increased by about 20 percent to allow for waste, loss, and pilferage.

As a rough check, approximately 100 pounds of nails are required per 4 MFBM or per 4 MSF of lumber or plywood involved, and no additional allowance for waste is necessary.

6.1.4 Material Suppliers or Subcontractors

Timber trusses and laminated wood beams are often included in the carpentry section of the estimate. These materials are usually quoted F.O.B. the job by prefabricated timber products suppliers.

In such cases, the estimator should set up separate sections on the estimate summary sheet to cover the purchase price for these materials. This procedure ensures that the quotations for these materials are not omitted from the estimate during the bid day rush, and permits the estimate sheet for the carpentry work to be completed and checked prior to bid day. The installation costs for trusses and laminated wood beams are carried in the estimate for the carpentry work. Lamella roofs are normally furnished and installed by timber engineering contractors on a sub-contract basis, and are usually covered by a separate section of the specifications. When lamella roofs are included in the carpentry section of the specifications, the estimator should set up a separate section for this work on the estimate summary sheet.

6.1.5 Hoisting

The rental charges for the truck cranes required to hoist or to erect carpentry materials are carried in the operating sections of the carpentry estimate. The item is included on the checklist to ensure that the estimator does not overlook this cost when preparing the estimate.

6.2   Unit Costs

In general, the unit costs derived for rough carpentry work exhibit greater variations from job to job, from day to day, or from one crew to another, than do those for any other type of work performed by the general contractor. Subtle differences in the framing details, in the amounts of work that can be done at any one time, and in the quality of the foremen, the crews, and the job superintendent, can result in cost variations as great as 100 percent.

When framing subcontractors are available, savings can usually be achieved by having them do this work. They are rough carpentry specialists, they do this work day after day, and their crews are highly skilled.

Unit costs for rough carpentry work are shown in Table 6.1. These

Table 6.1.   Unit Costs for Rough Carpentry

Activity	Labor $K_{6+2} = 1000$	Unit of Measure
Stud walls, H8–H10 (includes sills, plates, studs and blocking)		
2×4	$ 590	MFBM
2×6	$ 520	MFBM
Additional: Set anchor bolts for sills	$ 1.20	EA
Dry pack sills and set to grade	$ 2.20	LF
Floor joists (includes blocking)		
2×8	$ 430	MFBM
2×12	$ 400	MFBM
Roof joists (includes blocking)		
2×8	$ 450	MFBM
2×12	$ 430	MFBM
Ledgers (bolted to concrete walls)		
3×8	$ 700	MFBM
3×12	$ 650	MFBM
Set ledger bolts in wall forms	$ 1.50	EA
Furred Spaces (2×3 material)	$ 800	MFBM
2×3 Stripping (to ceiling joists)	$ 650	MFBM
2×4 Subfloor sleepers (metal clips additional)	$ 760	MFBM
2×6 Framed platforms	$ 720	MFBM
Fascia Boards		
2×8	$1000	MFBM
2×12	$ 860	MFBM
Eave framing (2×4, 2×6)	$ 660	MFBM

Table 6.1. (continued)

Activity	Labor $K_{6+2} = 1000$	Unit of Measure
Purlins		
2×3 bolted to steel @ 4′ centers.		
Includes drilling 3/16″ steel	$ 720	MFBM
2×4 screwed to steel @ 4′ centers	$ 430	MFBM
2×6 on wood (blocking additional)	$ 400	MFBM
2×6 T & G roof sheathing	$ 250	MFBM
Plywood (blocking additional)		
Subfloors ⅜″	$ 260	MSF
¾″	$ 280	MSF
1⅛″	$ 350	MSF
Shear walls ⅜″	$ 400	MSF
Roofs ½″	$ 220	MSF
¾″	$ 280	MSF
Exterior walls (⅜″ shingle panels		
8′0″ × 16″)	$ 750	MSF
4×2 Blocking for plywood	$ 790	MFBM
1×6 Diagonal sheathing (walls)	$ 500	MSF
Battens on wood walls		
1×4	$ 0.60	LF
2×4	$ 0.86	LF
Plaster grounds ¾×2	$ 0.60	LF

costs represent average values for first-class public work, and assume that the quantities of work that can be done at a given time are large enough to enable the crews to establish repetitive daily patterns, and that the rough carpentry can be prosecuted continuously for a period of several weeks. These unit costs include the costs for the saw men (table saw operators) and for the flat rack truck drivers if, and when, required.

In all cases, the unit costs for carpentry work are applied to the total MFBM or MFS of material purchased, including the allowances for waste.

Unit costs for finish carpentry work are shown in Table 6.2, and these costs are applicable to first-class public work performed by the general contractor's own forces.

The unit costs shown for installing unfinished wood cabinets are based on the areas of the front faces of the cabinets, and a single shelf is assumed to have a frontal area of 1 square foot per linear foot of shelf.

Table 6.2. Unit Costs for Finish Carpentry

Activity	Labor $K_{6+2} = 1000$	Unit of Measure
Doors		
Set hollow metal frames (to concrete)	$ 49	EA
Hang hollow metal doors complete	$ 21	EA
Set and trim wood jambs	$ 36	EA
Set rough wood bucks	$ 14	EA
Hang wood doors complete	$ 44	EA
Block out, set, and grout Rixson type hinges	$ 63	EA
Set frame and hang vault door (32′ × 78′)	$160	EA
Set frame and hang refrigerator door	$170	EA
Install panic hardware (additional)		
Wood jambs—per panic bar	$ 42	EA
Hollow metal frames—per panic bar	$ 63	EA
Install door closers	$ 22	EA
Install door stops		
Wall mounted	$ 5.50	EA
Floor mounted (concrete)	$ 10.00	EA
Install metal thresholds	$ 18.00	EA
Install name plates	$ 1.10	EA
Install kick or push plates	$ 6.30	EA
Install weather stripping (per door)	$ 11.00	EA
Install unfinished wood cabinets (per SF of face area)		
Open shelves	$ 1.20	SF
Typical cabinet installation	$ 1.55	SF
In congested areas	$ 3.00	SF
Install cabinet doors (additional)		
Small	$ 15	EA
Large	$ 20	EA
Install interior finish plywood		
1/4″ DF walls	$470	MSF
1/4″ DF ceilings	$870	MSF
1/4″ Birch walls	$890	MSF
Install toilet stalls		
Floor mounted	$ 46	EA
Ceiling hung	$ 96	EA
Install shoe mold or wood base	$ 0.60	LF
Install glass stop (¾″ × ¾″)	$ 0.54	LF
Install aluminum letters on concrete H6″–H8″	$ 11.50	EA
Install bronze plaques on concrete	$ 10.00	SF

Table 6.2.   (continued)

Activity	Labor $K_{6+2} = 1000$	Unit of Measure
Aluminum Sash (glazing additional)		
Set	$ 1.10	SF
Caulk (¼″ × ¼″—with gun)	$ 0.25	LF
Grout regelets	$ 90[a]	CF
Roll Up Doors (manually operated)[b]		
Set aluminum doors	$ 5.50	SF
Set steel doors	$ 4.00	SF

[a]Based on $K_F = 16.10$.
[b]The equipment cost for handling the doors is additional.

Experienced estimators often assign a lump sum cost per cabinet for installation, and the lump sum used is based on judgment and a careful consideration of each cabinet to be installed. In all cases, the cabinet sizes must be considered with respect to the room sizes and the doorway widths, and it may be necessary for some of the cabinets to be furnished in sections and assembled in place.

6.3   Heavy Timber Bents

The costs to frame and to erect two similar pieces of timber having identical cross sections will be essentially the same, even though their lengths may vary by 100 percent or more. Thus, the costs for heavy timber work cannot be based on the board feet involved, except in those cases where cost records from virtually identical work are available.

Structures such as heavy timber bents are framed and assembled, insofar as possible, on a level work area or platform which may consist of two or more rows of 10×10 or 12×12 timbers, blocked to grade. The platform should be as close to the point of erection as possible, and the material to be framed should be delivered nearby.

The timbers are placed on the platform with light equipment, such as an A-frame truck or a small rubber-tired backhoe, and are positioned and framed by carpenters using peaveys and cross-cut saws. The assembled sections are then erected with truck cranes.

Two carpenters, working on a platform, will position, set straight edges, and make one square cut on a 12×12 timber in about 0.33 hours, and this time should be increased by 50 percent if two cuts are required.

About two-thirds of the 0.33 hours will be spent handling and positioning the material, and this time will be essentially the same for timbers larger than 12×12. Thus, the estimated time required to make one square cut and to position a 16×16 timber would be about 2/3 × 0.33 + 1/3 × 0.33 × 256/144 = 0.42 hours for two carpenters.

In many cases some additional time is required for a laborer to furnish blocking and to assist the carpenters. The amount of laborer time allowed depends on the estimator's judgment. Indexed to K_{6+2} = 1000, a two-carpenter crew will cost 1000 ÷ 24 = \$41.67 per hour; this figure includes ⅔ hour of laborer time.

Two carpenters can make one square cut on an 8×8 timber in about 0.11 hours, and about one-half of this time is spent handling the material.

One carpenter can lay out and drill one hole through a 12×12 to 16×16 timber in about 0.08 hours, and install one bolt with washers in about the same amount of time.

Based on K_{6+2} = 1000, the cost to install 3×8 bolted bracing, complete, on horizontal bents is about \$350 per MFBM and the cost to install 3×8 bolted bracing on erected bents is about \$500 per MFBM.

The equipment rental costs to handle and to erect the bents depend on the size of the assembled units, the equipment that is available, and the number of assembled units that can be erected at one time.

Figure 6.1 illustrates a cost estimate for fabricating, erecting, and dismantling two timber falsework bents. The labor costs are indexed to K_{6+2} = 1000. It is assumed the fabricating platform is located near the point of erection, and that the materials are located close to the platform. The unit costs used are derived as follows:

Line 1.   Each platform timber will be blocked to grade at two locations. The unit cost to SS flat pads shown in Table 4.5 is used. In this case the pad loading will be small, but each pad must be blocked to a level grade.

Lines 2 and 3.   We assume a two-carpenter crew plus a small rubber-tired backhoe will set 16 pieces of timber on the pads or on the platform per hour. The bent timbers will be positioned by the carpenters as a part of the framing operation.

Line 4.   We assume the backhoe is located on the job. The second bent will be fabricated on top of the first one. Two on-off charges are required; one to set the timbers for the platform and the first bent, and another to set the timbers for the second bent.

Line 5.   Two cuts are required for each post; one to square the bottom, and another at the top to accommodate an assumed cross-fall at the cap. Thus, the time required per post will be about (0.33 × 2/3 + 0.33 × 1/3 ×

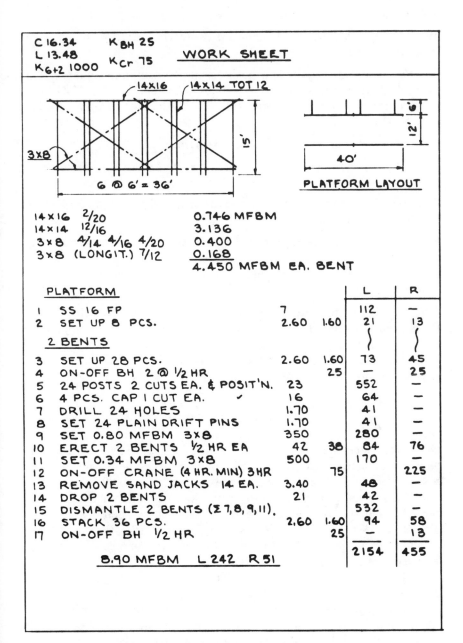

Figure 6.1 Cost estimate for timber bent.

196/144)1.5 = 0.55 hours. Assuming a two-carpenter crew, the cost per post will be 0.55 × 41.67 = $23.

Line 6.   The time required to position and to make one square cut on each 14×16 timber is estimated to be 0.33 × 2/3 + 0.33 × 1/3 × 224/144 = 0.39 hours. Assuming a two-carpenter crew, the cost per piece will be 0.39 × 41.67 = $16.

Lines 7 and 8.   0.08 × 41.67 × 0.5 = $1.70.

Line 10.   We assume a four-carpenter crew plus a truck crane will erect each bent in ½ hour.

Line 13.   We assume the posts are set on sand jacks, and that one laborer releases four jacks per hour.

Line 14.   After the sand jacks have been released, the falsework supported by the bents will be removed laterally, and the bents will be toppled. We allow an additional ¼ hour for a four-carpenter crew to topple each bent.

Line 15.   The bents were framed on a platform, but will be dismantled on the ground. This requires more time, and we therefore use the total costs for lines 7, 8, 9, and 11. The inclusion of line 7 in this total as well as the inclusion of line 11 at full cost compensates for the extra time required to release the drift-pinned connections between the caps and the posts.

6.4   Miscellaneous Carpentry Costs

Galvanized steel bench supports, made of ⅜″ × 2½″ material, were bolted to a concrete slab with three ½-inch expansion bolts each, at a unit cost for labor of $29 per bracket, indexed to $K_{6+2} = 1000$.

85 MFBM of 2×6 seat boards were installed on brackets at 4′-6″ centers, using two ⅜″ bolts per board per bracket, at a unit cost for labor of $360 per MFBM. This unit cost was achieved by a skilled and well organized crew. Small quantities of seat boards will be installed for a unit labor cost of about $420 per MFBM.

Roof vents, averaging 30 inches square, were framed, complete, at a unit cost for labor of $23 each, indexed to $K_{6+2} = 1000$.

Temporary job offices and sheds, made of prefabricated panels, will be erected and removed at unit labor costs ranging from about $4.00 to $6.00 per square foot of floor area, indexed to $K_{6+2} = 1000$. The higher unit cost applies when interior doors and partitions are required.

Rough 3×8 caps, 6 feet long, were bolted to steel shoring frames with eight ⅜″ × 2½″ lag screws, at a unit cost for labor of $2.90 each or $242 per MFBM, indexed to $K_{6+2} = 1000$.

7

Earthwork

Whenever earth is excavated, three basic operations are normally involved: the material must be excavated, it must be hauled, and it must be placed somewhere. Grades must always be made, and when embankments are to be constructed, the material must be watered and compacted.

On building work, the estimator decides what must be done, assumes a reasonable value for the side slopes, takes off the quantities, estimates their costs, and enters a lump sum cost for the earthwork on the estimate summary sheet.

The estimating of earthwork costs on unit price contracts requires more care, since only those quantities within the specified pay limits will be paid for. The take-off, therefore, must include both a check of the quantities shown on the engineer's estimate, which will be based on the pay limits, and a determination of the probable quantities to be handled, based on reasonable side slopes. In addition, on unit price contracts, the costs of certain additional work may be specified to be included in the price bid for a particular item.

Job scheduling sometimes requires material to be temporarily stockpiled. In such cases, two excavation operations are required, even though only one will be included in the pay quantities.

The water table must be below the bottom of the proposed excavation, and the costs to control ground water or to lower the water table, when necessary, are carried as a separate section of the cost estimate for the earthwork item.

When well point systems are necessary, the design of the system is furnished by the suppliers of the required equipment. The suppliers will quote rental rates for the equipment and will furnish cost estimates covering the installation, operation, maintenance, and removal of the system. The installation of the system is supervised by the supplier, but the labor and equipment required to install and to remove it are furnished by the general contractor, who will also operate and maintain the system.

The cost to excavate material depends on its physical properties, the quantities to be excavated, the equipment used, the skill of the equipment operators, the depth of cut, the size of the work area in relation to the equipment used, and the quality of the supervision.

The optimum depth of cut for a digging bucket is that which will permit the bucket to be fully loaded in one pass, and will vary with the size of the bucket and the hardness of the material being excavated. Scraper hauling units load best when making a 4 to 6 inch cut. Thus, a scraper with a 10-foot cutting edge and a capacity of 25 bank yards will require a minimum loading distance of about

$$\frac{25 \times 27}{0.42 \times 10} = 161 \text{ feet.}$$

7.1   Physical Properties of the Material

The type of material to be excavated is a basic cost factor, since soft materials load faster and swell less. Earthwork contractors become very proficient at moving those soils found in their particular areas, but extreme care must be exercised when estimating earthwork costs in areas where one is unfamiliar with the soils, and when large quantities are involved.

7.1.1   Density

The unit weight of the typical, in-place bank material must be considered since, in some cases, the amount of material that can be loaded or hauled per cycle will be governed by the weight instead of by the volume. In addition, the densities of the bank and of the compacted embankment materials are used to determine the grading factor, which is discussed in the following section.

The density of the bank material is determined by laboratory tests which give the dry density of the material and the percentage, by weight, of retained moisture. The in-place unit weight of the bank material will vary with the gradation of the material, the amount of moisture retained in it, and its natural relative compaction. The approximate in-place unit weights for commonly encountered materials are:

Material	Weight in pounds per bank cubic yard
Silty Sand	2700
Sand	3100
Earth	3100
Gravel	3400
Clay	3400

The percentage of retained moisture in the bank material and the percentage of moisture required for maximum density are both considered when estimating the amount of added water required for compaction.

7.1.2   Shrinkage and the Grading Factor

Material that has been compacted to 90 percent of its maximum density is said to have a relative compaction of 90 percent. Current specifications normally require a relative compaction of 90 percent for embankments, and the resulting dry density is usually greater than the dry density of the bank material. In such cases, more than 1 cubic yard of cut will be required to make 1 cubic yard of fill, and the material is said to shrink. This relationship, for a given material, is expressed by the grading factor, and in all cases

$$C = \frac{E}{G} \qquad (7.1)$$

where

G = the grading factor.
C = the total volume of cut required to make embankment E.
E = the volume of embankment.

The grading factor is determined as follows:

$$G = \frac{D_B}{D_E} \qquad (7.2)$$

where

D_B = the dry, in-place density of the bank material.
D_E = the average dry density of the embankment material, on completion.

Assuming a specified relative compaction for the embankment of 90 percent, D_E, in equation 7.2, will range from 90 to 100 percent of the maximum density for the material.

The average density of the completed embankment cannot be precisely determined in advance. Field tests for embankment densities are made in the top 1.5 feet of the embankment as the work progresses, and additional compaction is attained as loaded rubber-tired equipment rolls over the tested layer while placing the layers above.

Grading factors are given on the job plans, and the values used will depend on the judgment and the experience of the engineer. Typical values for the grading factor will range from 0.75 to 0.90, when embankments are made of earth to a relative compaction of 90 percent.

The grading factor is widely used by highway departments and contractors to adjust earthwork volumes. Many soils engineers and civil engineers in private practice, however, use a shrinkage factor that is the reciprocal of G. In such cases, equation 7.1 becomes

$$C = ES$$

where

$$S = \text{the shrinkage factor} = \frac{1}{G}.$$

When embankments are made of a mixture of blasted rock or boulders and earth, the average density of the embankment cannot be determined with much accuracy, and the grading factor will usually be greater than 1 and at best, be an educated guess.

When high fills are constructed, the subgrade may subside under the added weight, and additional cut will be required to complete the fill. An allowance for subsidence is normally included in the embankment quantities given on the road profiles, and the amount of subsidence assumed will usually range from 0.15 to 0.30 feet. It is important to note that, while the allowance for subsidence increases the cross sectional areas of the embankment sections, subsidence does not affect the grading factor, which relates to shrinkage.

Large areas requiring shallow cuts and fills are often graded with wheel tractor-scrapers. Such equipment will compact the original ground surface during the grading operation, and the resulting subsidence may be as much as 0.4 foot. Since shallow cuts and fills over a large area are involved, this subsidence will cause a significant decrease in the amount of excavation otherwise available and will increase the amount of fill required, in those cases where the grading plane was established without considering the subsidence.

To obtain an earthwork balance, in such cases, the grading plane is often established so that the volume of excavation is as much as 50 percent greater than the volume of fill, and this percentage covers the combined effects of shrinkage and subsidence. The proper selection of the percentage to be allowed requires past job experience involving similar work and similar soils.

7.1.3   Swell and the Load Factor

When bank material is excavated and loaded, voids are created and the material is said to swell. This relationship is expressed by the load factor as follows:

$$L \times F = B \tag{7.3}$$

where

F = the load factor.
B = the number of bank yards excavated.
L = the number of loose (excavated) yards resulting from the excavation of B bank yards

and

$$F = \frac{D_L}{D_B} \tag{7.4}$$

where

D_L = the density of the loose material.
D_B = the density of the bank material.

Equation 7.4 may be restated as

$$F = \frac{1}{1 + \dfrac{S}{100}} \tag{7.5}$$

where

S = the percentage of swell, and from equation 7.5,

we have

$$S = \left(\frac{1}{F} - 1\right) \times 100 \tag{7.6}$$

Approximate load factors for commonly encountered materials are:

Material	Load Factor
Sand	0.90
Gravel	0.90
Earth	0.80
Clay	0.78
Hardpan	0.65
Blasted rock	0.60
Sandstone	0.60
Shale	0.60
Broken concrete	0.58
Stockpiled material	1.00

The load factor may be used to approximate the number of bank yards contained in a loader scraper, truck, or bucket. Thus, if the load factor were 0.80, the number of bank yards contained in a 5 CY bucket would be $0.80 \times 5 = 4.0$ CY. In practice, estimating the number of bank yards carried per load is not that simple.

The load factor, or swell, taken from a table is an approximation, at best, and will vary in the same material, depending on the method of excavating and loading. It may also vary by as much as 6 or 8 percent for similar pieces of equipment built by different manufacturers. In addition, earthmoving equipment may be rated at struck and at heaped capacity, and sound judgment is required to select that capacity to which a load factor should be applied.

On unit price contracts, an error in the load factor is more serious than an error in the grading factor. If the assumed grading factor were 5 percent too high, 5 percent more excavation would be required, and the haul lengths would change slightly. The job would end up short of dirt, and the additional excavation required would be paid for under a contract change order. The contractor would, however, receive his bid price for each yard excavated under the item, as well as his negotiated price for each yard handled under the change order.

If, on the other hand, the assumed load factor were 5 percent too high, 5 percent more equipment cycles would be required to handle the contract amount of excavation, and no payment would be made for the additional work, since the number of cubic yards excavated would not have changed.

In practice, the number of bank yards carried per load is based on experience and past job records for work done with similar equipment in similar types of soil, and the job records will include actual load counts relating to surveyed amounts of yardage.

7.1.4 Moisture Required for Compaction

The percentage of moisture, by weight, required for maximum density is determined from laboratory tests. Points are plotted to produce a curve having dry densities for ordinates and percentages of moisture for abscissas. As increasing amounts of moisture are added to successive samples, the compacted densities increase until maximum density is reached, after which the densities decrease as more moisture is added.

The moisture-density curve is plotted so that the distance representing a difference of 1 pound of density on the vertical scale is the same as the distance representing a difference of 1 percent of moisture on the horizontal scale. The shape of the resulting curve has an important bearing on the relative ease of compacting the material on the job. The slope of the curve on either side of the maximum point will be 45 degrees or flatter for sandy soils, and their specified compaction can be achieved over a relatively wide range of moisture content.

The same slopes will usually be steeper than 45 degrees for silty soils, and quite steep for soils containing 50 percent or more of material passing the number 200 sieve. The specified compaction for such soils can only be attained when the moisture content is very close to the optimum, and it is not uncommon for the compacted densities of silty soils to decrease from 7 to 10 pounds per cubic foot for a 1 percent increase above the optimum moisture content.

Since the amount of moisture cannot be precisely controlled on the job, it is clear that soils having steep moisture-density curves will present more problems during the compaction operation. Discs are usually required to facilitate the compaction of such materials. Small amounts of additional water can easily be mixed with material layers that are too dry, and when a layer is too wet, the discs will aerate the material and shorten the time required for the drying out process.

7.1.5 Gradation

Well graded soils compact more easily than poorly graded ones. On rare occasions, a poorly graded, silty sand may be encountered, which cannot be compacted, under job conditions, to a relative compaction of 90 percent. Such soils may have a relatively high percentage of material retained between the No. 100 and the No. 140 sieves, and under a microscope this fraction will be roughly spherical in shape. In such cases, it may be necessary to blend a small amount of sand with the bank material to achieve the required compaction.

7.2 Earthmoving Equipment

Production rates derived for earthmoving equipment are often based on the assumption that the equipment will work steadily, with no delays. Such rates are said to be based on a 60-minute hour. On all jobs, minor delays are inevitable, and, for estimating purposes, effective rates based on 45- to 55-minute working hours are used. A production rate of 500 cubic yards per 60-minute hour will result in an effective rate of

$$500 \times \frac{50}{60} = 417 \text{ CY/hr, based on a 50-minute hour.}$$

The minutes per working hour, used by the contractor, will depend on his past cost records for particular types of equipment and on the anticipated job conditions. Effective production rates for track-type equipment are often based on 50- to 55-minute hours, and 45- to 50-minute hours are often used for rubber-tired equipment. Some successful trucking contractors use a 55-minute hour for on-highway hauling.

7.2.1 Bulldozers

Bulldozers are most effective when the haul lengths are 300 feet or less. As indicated by the production rates below, a difference of 50 feet in the average haul length has a significant effect on the production achieved.

Approximate Bulldozer Production
Track type, power shift machine, 300 flywheel horsepower, straight blade

Haul distance (feet)	Base Production Rate (LCY per 60 minute hour)
50	800
100	525
150	375
200	300
300	200
400	175
500	150

Production rates are, in general, proportional to the horsepower of the machine used. The above rates assume an average operator, working on a level grade in average earth. For other conditions, the following approximate correction factors should be applied.

Condition	Factor
Loose stockpiled material	1.2
Ripped rock	0.7
Dry sand or silt, mud, or wet clay	0.8
Dozing up a 20% grade	0.6
Dozing down a 20% grade	1.2
Direct drive transmission	0.8
U Blade	1.2
Angle blade	0.65

Production rates will vary for different makes and models of equipment. When possible, rates from past jobs having similar conditions should be used. When such data are not available, production rates for a particular machine should be obtained from the manufacturer.*

Based on the approximate production rates shown above, and assuming the following conditions:

Bulldozer	250 HP, direct drive
Material	Ripped rock
Grade	Level
Haul distance	200 feet, one way
Working hour	50 minutes

the estimated production rate would be

$$300 \times \frac{250}{300} \times 0.8 \times 0.7 \times \frac{50}{60} = 117 \text{ LCY/hr.}$$

Assuming a load factor, for ripped rock, of 0.60, this is equivalent to

$$117 \times 0.60 = 70 \text{ BCY/hr.}$$

*Caterpillar Tractor Co., *Fundamentals of Earthmoving* and the *Caterpillar Performance Handbook* deserve a place in every contractor's library.

Figure 7A Caterpillar D9 bulldozer, Parallelogram-type ripper. (Meadow Construction Company, Bakersfield, California).

7.2.2   Rippers*

The production rate for a ripper will depend on the tractor horsepower and tractive capability, and on the degree of consolidation, or rippability, of the material.

Clays, shales, cemented gravel, and hardpan are easily ripped. Rock formations may be sedimentary, igneous, or metamorphic, and of these, sedimentary rocks are most easily ripped. The rippability of sedimentary rock varies with its hardness, the degree of decomposition, the thickness of the strata, and the amount of fracturing. Unfractured strata more than about 12 inches thick are difficult to rip. In general, igneous and metamorphic rocks can only be ripped when they are fractured or decomposed.

7.2.2.1   REFRACTION SEISMIC ANALYSIS.   On highway work, cuts must sometimes be made in areas where the topsoil covers one or more layers of rock. In such cases, test boring data on the job plans will show the depth to the rock in the cut areas. When the subgrade for the proposed excavation is below the rock layers, the rippability of the rock may be determined with reasonable accuracy by refraction seismic analysis.

Such analysis is based on the speed with which seismic waves travel through the different kinds of subsurface materials; these speeds range

---

*Most of the material in this section has been taken from the following two Caterpillar Tractor Company publications:

1. *Handbook of Ripping,* Sixth Edition.
2. *Caterpillar Performance Handbook,* Edition 8.

from 1,000 feet per second through loose soil to about 20,000 feet per second through hard, solid rock.

The seismic equipment consists of a geophone connected to an electric cable to which steel striking plates are attached. The striking plates rest on the ground, and are spaced along the cable at 10-foot intervals. The seismic waves are produced by striking the plates, in succession, with a sledge hammer. The time intervals between the hammer blows and the arrival of the waves at the geophone are recorded, and an analysis of these data indicates the depths, thicknesses, and rippability of the various subsurface layers.

Seismic surveys are typically made by geological exploration firms prior to the bid date, their reports are sold to the prospective bidders, and the price paid depends on the number of contractors using the information. Assuming a 410 flywheel horsepower tractor equipped with a single shank, parallelogram-type ripper, rock is generally considered rippable when the seismic wave velocities are within the following limits:

Igneous rock	3000–7500 feet/second
Metamorphic rock	3000–7500 feet/second
Sedimentary rock	3000–8500 feet/second

Ripper tooth penetration is essential if rock is to be ripped. Although seismic velocities for sedimentary rock may indicate probable rippability, the material may not be ripped effectively if the fractures and bedding joints do not allow tooth penetration.

7.2.2.2   RIPPER PRODUCTION.   Accurate estimates of ripper production in relation to seismic wave velocities can only be made by contractors who have production records and seismic data covering past work with their equipment, in their area. The Caterpillar Tractor Co. has developed the following data, however, with the proviso that they be used with discretion and in conjunction with job experience in ripping work.

Velocity of seismic waves (feet per second)	Production (bank yards per 60 minute hour)
3000	1500–3000
4000	1000–2250
5000	500–1700
6000	250–1200

These data apply to a 410 flywheel horsepower tractor equipped with a single shank parallelogram-type ripper, and, in general, ripping production is proportional to the tractor horsepower and tractive ability. The upper production rates shown are only applicable when ripping conditions are ideal; under adverse conditions, the lower limits should be used.

When the production rate falls below 150 to 200 bank yards per hour, production may sometimes be increased by as much as three or four times if a second tractor is used as a pusher.

Generally, the most efficient ripping is done at speeds of 1 to 1.5 miles per hour. When working in soft material, it is preferable to use 2 or 3 teeth, rather than increase the speed.

When ripping soft material, and when the passes are longer than about 125 feet, maximum production is achieved when successive passes are made in opposite directions. Assuming a ripping speed of 1 mile per hour (88 feet per minute), and allowing for a 0.5 minute delay per pass at the turns, an estimate of the probable production, in such cases, may be made as follows:

$$S = \frac{88L}{L + 44} \tag{7.7}$$

where

S = effective travel speed in feet per minute.
L = length of pass in feet.

and

$$P = \frac{60 \times S \times W \times D \times K}{27} \tag{7.8}$$

where

P = production rate in bank yards per 60-minute hour.
W = width loosened per pass, in feet.
D = depth of penetration in feet.
K = a constant based on job experience. The Caterpillar
Tractor Co. recommends K = 0.80 to 0.90.

Assuming the following conditions:

270 horsepower ripper with 1 tooth
6 feet between passes
2.5 feet of penetration
250 foot pass lengths

we have

$$S = \frac{88 \times 250}{294} = 75 \text{ feet per minute}$$

and
$$P = \frac{60 \times 75 \times 6 \times 2.5 \times 0.80}{27} = 2000 \text{ bank yards per 60-minute hour}$$

When the passes are shorter than about 125 feet, or when required by job conditions, passes are made in one direction and the tractor backs up after each pass. Assuming a ripping speed of 88 feet per minute, and a reverse speed of 250 feet per minute, the effective speed for the ripping pass is

$$S = \frac{L}{\dfrac{L}{88} + \dfrac{L}{250}} = 65 \text{ feet per minute}$$

and this value should be used directly in equation 7.8 to estimate production.

Heavy ripping in rock results in high repair costs. Ripper tips are expensive, and might have a life of about 1 month to as little as 30 minutes, under the worst conditions. Tip life, under severe conditions, is sometimes estimated as 10 hours. However, as many as 1,000 hours have been obtained in shale. Due to accelerated wear, the hourly costs for rippers working in heavy rock will be about two times the normal cost for the equipment.

7.2.3   Track Loaders and Wheel Loaders

Loader buckets are rated at heaped capacity, which assumes 2:1 slopes for the material above the struck capacity line, when this line is parallel to the ground. Under job conditions, the struck line will not be exactly parallel to the ground, the bucket will swing and bob, and a carry or spill factor is applied to the rated capacity when estimating loader production.

7.2.3.1   BANK YARDS PER CYCLE.   The bank yards carried in the bucket, per cycle, are

$$B = R \times L \times C$$

Figure 7B Track-type loader, (The Caterpillar Tractor Co.).

where

B = bank yards in the bucket.
R = rated bucket capacity.
L = load factor.
C = carry factor.

The Caterpillar Tractor Co.* recommends the following values for the carry factor.

Track Type Loader Carry Factors

Material	Factor
Uniform aggregates	0.85–0.90
Mixed moist aggregates	0.95–1.00
Moist loam	1.00–1.10
Soil with boulders and roots	0.80–1.00
Cemented materials	0.85–0.95

*Caterpillar Tractor Co., Fundamentals of Earthmoving, July 1975.

Figure 7C Loader-backhoe.

Wheel Type Loader Carry Factors

Material	Factor
Uniform aggregates	
Sand	0.95–1.00
¼"	0.85–0.90
⅝"	0.90–0.95
1" and over	0.85–0.90
Mixed moist aggregates	0.95–1.00
Blasted material	
Well blasted	0.80–0.85
Average	0.75–0.80
Slabs or blocks	0.60–0.65

When loaders are used in hard or rocky material, when the depth of cut is far from optimum, or when top production rates are required, the excavation is made with bulldozers, and the loaders work from stockpiles. As noted previously, the load factor for stockpiled materials is assumed to be 1.00, for estimating purposes.

Aggregates are usually paid for by the ton, whereas buckets are rated in cubic yards. The approximate densities for stockpiled aggregates are

Stockpiled material	Density (tons per cubic yard)
Dry sand	1.35
Moist sand	1.45
Dry gravel	1.32
Dry sand and gravel	1.50
Wet sand and gravel	1.69

Figure 7D Caterpillar wheel loader, (5M Company, Bakersfield, California).

When loaders are used to load trucks, the truck and the loader capacities should be balanced, when possible, so that the trucks may be fully loaded with a whole number of passes.

7.2.3.2 CYCLE TIMES.   The base cycle time for loaders is the time required to fill the bucket, swing 60 to 90 degrees, dump the material, and return to the bank or pile. For estimating purposes, the base cycle times, based on a 60-minute hour, will be about as follows:

Track type loaders   0.6 minutes
Wheel type loaders   0.5 minutes

When a swing of 180 degrees is required, these base cycle times increase about 33 percent.

When the loader must carry the material to a dumping site, the travel time must be added to the base cycle time above. Average travel speeds for loaders, based on a 60-minute hour, are about as follows:

Loader	One-way Distance (feet)	Speed (feet per minute)
Track type	50	200
	100	330
	200 plus	500
Wheel type	up to 500	370
	500 plus	800

Figure 7E Caterpillar motor grader, (J. L. Denio Inc., Bakersfield, California).

To achieve 800 feet per minute, the wheel loader must have a smooth haul road.

7.2.4 Motor Graders

The production for motor graders or blades is, in most cases, related to the areas or distances covered, rather than to cubic yards of material. Thus, unit costs for motor grader work are normally based on square yards, or stations, per hour.

$$P_A = \frac{S \times 5280 \times W}{9 \times N} = \frac{587 \times S \times W}{N} \qquad (7.9)$$

where

P_A = production rate in square yards per 60-minute hour.
S = travel speed in miles per hour.
W = effective blade width in feet.
N = number of passes required.

or

$$P_S = \frac{S \times 52.80}{N} \qquad (7.10)$$

where

P_S = production rate in stations per 60-minute hour.

For estimating purposes, an average speed of 4 to 4.5 miles per hour is often used, and the effective blade width is taken as 10 feet for finish work and 12 feet for rough work, such as mixing water with fill material during the compaction of embankments.

7.2.4.1 MISCELLANEOUS MOTOR GRADER COSTS. Roadside ditches, approximately 2 feet deep, with 4:1 side slopes, were excavated, complete, with a motor grader. Indexed to $K_L = 13.48$, and to an hourly rental rate of $39 for the motor grader, the unit costs were

Labor $0.45 per cubic yard
Rental $1.30 per cubic yard.

These unit costs do not include disposing of the excavated material.

A motor grader was used to excavate, backfill, and make the final clean up for 200 cubic yards of curb and gutter. The curb and gutter section contained 0.0617 cubic yards of concrete per linear foot, and the total length was 32.4 stations. Indexed to $K_L = 13.48$, and to an hourly rental rate of $39 for the motor grader, the unit costs for this work were

Labor $ 0.55 per cubic yard of concrete
  or $ 3.40 per station
Rental $ 2.70 per cubic yard of concrete
  or $16.70 per station.

7.2.5   Tractor-Scrapers

Scrapers may be drawn by track-type tractors, or by rubber-tired tractors (wheel tractors). Wheel tractor-scrapers are discussed, with examples, in Chapter 9. The same principles apply to tractor-scrapers, but the maximum travel speed will be about 8 miles per hour on well-maintained haul roads, and as low as about 4 miles per hour when haul roads are poor.

The loading time for tractor-scrapers is often taken at about 1 minute for small scrapers less than 12 cubic yards in capacity, and at 1.5 minutes for larger scrapers. For maximum production, a push-tractor may be required.

7.2.6   Compaction Equipment

The production rate for compacting equipment may be estimated as follows:

$$P = \frac{S \times 5280 \times W \times T}{27 \times N \times G} = \frac{196\, S \times W \times T}{N \times G} \qquad (7.11)$$

Figure 7F Caterpillar D8 tractor with 12 CY scraper.

where

P = production rate in bank yards compacted per 60-minute hour.
S = travel speed in miles per hour.
W = width of roller in feet.
T = thickness of the compacted lift in feet.
N = number of passes required.
G = grading factor.

The average travel speed will be about as follows:

Vibratory rollers	2 MPH
Self-propelled sheepsfoot	6 MPH
Self-propelled pneumatic	7 MPH
Sheepsfoot towed by a track type tractor	3–4 MPH

Experience with the soils in question is required to estimate the number of passes necessary to achieve the specified compaction, but in most cases, from 4 to 6 passes are sufficient. As previously discussed, the ease of compaction varies with the gradation of the soil, its moisture content,

Figure 7G Vibratory sheepsfoot roller, (Southwest Welding and Manufacturing Co., Alhambra, California).

and the uniformity of the soil-moisture mixture. In most cases, soils compact more easily, on the job, when the moisture content is about 1 percent greater than the optimum.

Sheepsfoot rollers work well in sandy or in clay soils. Since they compact from the bottom up, the compacted thickness of the individual layers is limited to the length of the feet, which will range from 8 to 9 inches.

Figure 7H Caterpillar 815 compactor, (The Caterpillar Tractor Co.).

Figure 7I Double 5 × 5 Sheepsfoot Roller.

A double 5×5 sheepsfoot roller consists of two rollers, side by side, each 5 feet in diameter by 5 feet wide, and is normally towed by a track-type tractor. Such a roller has an effective width of about 11 feet, and will compact sandy loam material to a relative compaction of 90 percent at an effective rate of about 700 bank yards per hour, when working on embankments placed with wheel tractor-scrapers.

Vibratory rollers work best in granular materials, and are effective to a depth of 2 feet or more. A single vibratory sheepsfoot roller 5 feet in diameter by 6 feet wide will compact sandy loam material at an effective rate of about 400 bank yards per hour, and no assistance will be required from the hauling equipment.

7.2.7   Trucks

When large amounts of material are to be hauled over public roads, or when prolonged grades are involved, test runs are made with loaded and empty trucks of the same type that will be used on the proposed work, to determine the average travel speeds. Typically, trucks will travel at an average speed of 40 miles per hour on good county roads, and at an average speed of about 26 miles per hour in urban areas where traffic is light.

Table 7.1. Estimated Dragline Production (BCY per 60-minute hour)

Material Class	Dragline Bucket Size (in cubic yards)												
	¾	1	1¼	1½	1¾	2	2½	3	3¼	4	4½	5	6
Moist loam or sandy clay	130	160	195	220	245	265	305	350	390	465	505	540	610
Sand and gravel	125	155	185	210	235	255	295	340	380	455	495	530	600
Good, common earth	105	135	165	190	210	230	265	305	340	375	410	445	510
Clay, hard and tough	90	110	135	160	180	195	230	270	305	340	375	410	475
Clay, wet and sticky	55	75	95	110	130	145	175	210	240	270	300	330	385

Note: These data assume average job conditions, a 90-degree swing, optimum depth of cut, and that the material is loaded into hauling units.

Table 7.2. Standard Boom Lengths for Draglines

Bucket size (CY)	¾	1	1½	2	3	4	5
Boom length (feet)	35–50	40–55	50–70	50–90	60–110	70–120	70–130

Although on-highway trucks may reach a top speed of 40 miles per hour on well maintained, dirt haul roads, the average speed, in most cases, will be about 27 miles per hour.

Whenever trucks are used, a load counter should keep a log of the truck cycle times; when large amounts of material are being hauled from off-job locations, a truck foreman should be used to keep the operation running efficiently.

Good judgment is required when hauls must be made across railroad tracks. In all cases, data on the anticipated train traffic during the haul period must be obtained and carefully considered. When large quantities of material must cross heavily traveled tracks, temporary haul bridges or conveyor belt systems may be necessary.

7.2.8 Draglines*

Dragline production may be roughly estimated from the data shown below, which assume that mass excavation is being performed, that close adherence to lines and grade is not required, and that the boom length is at or near the lower end of the ranges shown in Table 7.2.

Assume the following conditions:

Bucket size	1 CY
Swing	120 degrees
Depth of cut	8 feet
Material	Sand
Working hour	50 minutes

The optimum depth of cut, from Table 7.3, is 6.6 feet, and the job cut is

$$\frac{8}{6.6} \times 100 = 121 \text{ percent of optimum.}$$

---

*Tables 7.1 through 7.4 are copyrighted material taken from Technical Bulletin No. 4 of the Power Crane and Shovel Association, and are reproduced here with their permission.

Table 7.3. Optimum Depths of Cut in Feet for Short Boom Dragline Performance

Class of Material:	Bucket Capacity (in cubic yards)												
	⅜	½	¾	1	1¼	1½	1¾	2	2½	3	3½	4	5
Light moist clay or loam	5.0	5.5	6.0	6.6	7.0	7.4	7.7	8.0	8.5	9.0	9.5	10.0	11.0
Sand or gravel	5.0	5.5	6.0	6.6	7.0	7.4	7.7	8.0	8.5	9.0	9.5	10.0	11.0
Good common earth	6.0	6.7	7.4	8.0	8.5	9.0	9.5	9.9	10.5	11.0	11.5	12.0	13.0
Clay, hard, tough	7.3	8.0	8.7	9.3	10.0	10.7	11.3	11.8	12.3	12.8	13.3	13.8	14.3
Clay, wet, sticky	7.3	8.0	8.7	9.3	10.0	10.7	11.3	11.8	12.3	12.8	13.3	13.8	14.3

Table 7.4. Correction Factors for the Effects of Depth of Cut and Angle of Swing on Dragline Production

Depth of Cut (in % of optimum)	Angle of Swing (in degrees)							
	30	45	60	75	90	120	150	180
20	1.06	0.99	0.94	0.90	0.87	0.81	0.75	0.70
40	1.17	1.08	1.02	0.97	0.93	0.85	0.78	0.72
60	1.25	1.13	1.06	1.01	0.97	0.88	0.80	0.74
80	1.29	1.17	1.09	1.04	0.99	0.90	0.82	0.76
100	1.32	1.19	1.11	1.05	1.00	0.91	0.83	0.77
120	1.29	1.17	1.09	1.03	0.98	0.90	0.82	0.76
140	1.25	1.14	1.06	1.00	0.96	0.88	0.81	0.75
160	1.20	1.10	1.02	0.97	0.93	0.85	0.79	0.73
180	1.15	1.05	0.98	0.94	0.90	0.82	0.76	0.71
200	1.10	1.00	0.94	0.90	0.87	0.79	0.73	0.69

From Table 7.1, the estimated production rate is 155 BCY per 60-minute hour, and from Table 7.3, the correction factor is 0.90. The effective production rate is estimated to be

$$155 \times 0.90 \times \frac{50}{60} = 116 \text{ BCY per hour.}$$

8

Structure Excavation and Structure Backfill

Structure excavation and structure backfill are taken off together, but are normally bid as two separate job items. The quantities involved are relatively small; the upper vertical pay limit for structure excavation is frequently the lower pay limit for a mass excavation item, and the upper limit for the structure backfill may be the sub-grade for an embankment.

8.1 Structure Excavation

Typically, the excavated material is stockpiled at the site, and is subsequently used for structure backfill. After the backfill is completed, any excess material is incorporated into an adjacent fill at no additional cost, or is spread out and "lost" during the final cleanup operation.

Structure excavation is normally done with equipment to within about ±0.1 foot of finish grade, and the fine grade is made by hand. When concrete structures are involved, the area to be fine graded is taken off and priced with the concrete; typical costs for such fine grading are shown in the Unit Cost Tables in Chapters 4 and 5. The fine grading costs relating to pipe excavation are included in the unit costs for laying the pipe.

8.1.1 Hand Excavation

A detailed analysis of hand excavation costs is seldom warranted, since the total cost for hand excavation normally represents an insignificant portion of the overall job cost.

In sandy soil, free from rocks, where the material can be cast to the side, one laborer will excavate about 1 CY per hour. If picks and wheelbarrows are required, this basic cost increases.

When deep underpinning pits are excavated by hand, a power winch and buckets are required. A three man crew is used with one man in the pit, one man at the head of the pit, and one man on the winch. In easy digging such a crew will excavate from 0.4 to 0.6 CY per hour when working in a pit four feet square.

8.1.2 Unit Costs

Unit costs for structure excavation are shown in Table 8.1. The unit costs for labor include the cost of a grade checker as well as the costs for that incidental hand excavation required to remove material from the bottom corners of the excavated sections.

The horizontal pay limits for pipe excavation are normally specified to be vertical planes 1 foot outside the neat lines of the pipe. In most cases trenches less than 5 feet deep are dug with vertical banks, and the volume of the excavation required will equal that shown on the engineer's estimate.

When estimating pipe excavation costs, the trench depths must be determined, since vertical, unshored banks are limited to 5 feet in height

Table 8.1.   Unit Costs for Structure Excavation

Type of Structure or Activity	Labor		Rental		Unit of Measure
	Unit Cost	Index	Unit Cost	Index	
Box culverts[a]	$0.90	$K_L = 13.48$	$2.65	$K_{Cr} = 75$	CY
Flat slab bridges[b]	$0.70	$K_L = 13.48$	$1.75	$K_{Cr} = 75$	CY
Girder bridges	$0.80	$K_L = 13.48$	$3.10	$K_{Cr} = 75$	CY
Pipe excavation[c]	$1.20	$K_L = 13.48$	$2.00	$K_{BH} = 25$	CY
Grade beam excavation	$1.10	$K_L = 13.48$	$1.70	$K_{BH} = 25$	CY

[a]These costs apply to box culverts located in existing channels where the bulk of the excavation is for the wingwalls.
[b]Similar to bridge shown on Figure 4.17.
[c]Depth about 5 feet, and work area relatively flat.
Note: These unit costs are applicable to work in ordinary earth, and include the costs for checking grades. The unit costs should be applied to the true quantity of excavation, which may be greater than that shown on the engineer's estimate. On-and-off charges for equipment are additional.

by safety regulations. On many jobs, profiles for the pipe work are not shown, but the engineer's estimate for the pipe excavation is tabulated for each run of pipe. In such cases the average trench depth is determined for each run of pipe as follows:

$$\bar{D} = \frac{27\,E}{LW}$$

where

$\bar{D}$ = average depth of trench in feet.
E = the engineer's estimated cubic yards for the run of pipe being considered.
L = the length of the run in feet.
W = the width between the horizontal pay limits in feet.

Typically, on freeway work, the average depth so determined will closely approximate the true average depth, and this can be verified, when necessary, by a review of the typical cross sections for the area under consideration.

When trenches are to be between about 5 and 8 feet in depth, the upper layer of material may be moved to one side with a bulldozer, and the bottom 5 feet excavated with vertical banks. During the backfill operation, the upper layer is replaced and compacted by the mass backfill method.

The unit costs for equipment, shown in Table 8.1, reflect the low production rates achieved in this type of excavation. Thus the effective production rate for box culvert excavation is 75/2.65 = 28 CY per hour, and is based on using a ¾ CY dragline for the box section, and a ¾ CY clamshell bucket for the headwalls. It is assumed the box culvert is in a natural channel, or in a channel excavated under a mass excavation item, and that only a small amount of excavation is required for the box section.

The low production rates for structure excavation result from the small amounts of excavation possible per set up, the necessity to closely control the lines and grade, and the inability of the equipment to set and bail.

The unit costs for girder bridges include the costs of excavating for the abutments and for the piers, and all unit costs shown assume that the excavated material is stockpiled at the site.

Well defined hand excavation such as that required for the retaining wall foundation keys, illustrated in Chapter 4, is taken off and priced with the related concrete. Incidental amounts of hand excavation are, however, required in connection with excavation done with machines.

When unit costs for structure excavation are derived from production rates for equipment, the labor cost for the grade checker is easily determined, and is applied to the total amount of machine excavation. The cost of the incidental hand excavation is, in such cases, based on a careful consideration of the typical excavation cross sections.

Thus, if a total of 1000 CY of structure excavation has been taken off, and if the estimator concludes that 5 percent of this material must be removed by hand, the unit cost for equipment would be applied to 950 CY, the unit cost for hand excavation would be applied to 50 CY, and the total costs for these two items would be entered in the rental and labor columns of the estimate sheet.

8.1.3 Footing Excavation with Trenching Machines

Wheel type trenching machines are frequently used to excavate for the grade beams in building foundation work. The grade beams are then poured neat, the structure backfill is eliminated, and savings are achieved.

When trench lengths are greater than 100 feet, trenches 20 inches wide by 3 feet deep can be dug in ordinary earth at a rate of about 280 LF per hour. Short trenches averaging 25 feet long can be dug at a rate of about 100 LF per hour. In all cases an allowance must be made for moving the machine on and off the job, and a grade checker will be required.

8.1.4 Sheeting and Shoring

Current safety regulations require trenches or banks greater than 5 feet in depth or height to be supported with sheeting, or laid back to a safe slope which must be no steeper than ¾ horizontal to 1 vertical.

Where large areas are to be excavated, and where space permits, it is usually cheaper to slope the banks than to shore them. When excavating for retaining walls, or for building foundations and basements, however, sophisticated shoring systems are frequently required to support the banks and adjacent roads or buildings. Such work is normally done on a sub-contract basis by pile driving or foundation contractors specializing in this type of construction.

On occasion, small amounts of short steel sheet piles, for temporary support, may be set and driven by the general contractor, using a truck crane equipped with hanging leads and a drop hammer. A temporary waling system is required and it may consist of a 4×12 sill at the bottom, 4×12 posts 5 feet high at 6 foot centers, and a 4×12 cap. The waling is braced with 2×8's and 2×4 stakes at 6 foot centers.

All the piles are set and braced against the wales before starting to drive. The entire row is then driven in a series of passes, and the driving depth is about 12 inches per pass. An inexperienced three-man crew will set and drive sheet piles 12 feet long in this manner at a rate of about 90 square feet per hour. Based on $K_{6+2} = 1000$, the wales will be set and stripped for about $450 per MFBM. In addition to the on and off charges for the crane and leads, about 6 hours will be required to rig up and dismantle the leads and hammer.

A pit, located in sand, 6 by 26 feet in plan and 12 feet deep was sheeted with 3×12 vertical timbers supported by three sets of 6×6 wales with 6×6 struts. The timber was set for $150 per MFBM, based on $K_{6+2} = 1000$. The excavation was made with a ¾ CY clamshell bucket at an effective rate of 8 CY per hour.

Banks 30 feet high were shored by using properly reinforced CIDH piles together with 2×12 horizontal sheeting. The 16-inch diameter piles were installed in a row on 3-foot centers. The pile tips extended 12 feet below the bottom of the proposed excavation, and the material was moist sand.

The excavation along the face of the piles was done with a ¾ CY clamshell bucket, and pieces of 2×12 sheeting approximately 2 feet long were placed horizontally, between and slightly behind the centerline of the piles, as the excavation progressed.

The excavation was done at an effective rate of 20 CY per hour, and the 2×12 material was installed for $1,730 per MFBM, based on $K_L = 13.48$. The piles were supported by two steel wales together with timber struts.

8.2   Structure Backfill

Structure backfill on first-class work must be compacted to a relative compaction of 90 or 95 percent, and is a costly item. Unfortunately, some smaller agencies specify such compaction, but do not take tests or require it in the field. Small jobs have been lost when the contractor bid the price to place structure backfill as specified, on projects where the competition knew the specified compaction would not be required.

The outside horizontal pay limits for structure backfill are the same as those for structure excavation, and the true volume placed will often amount to two or three times the amount shown on the engineer's estimate.

Public agencies frequently require sand to be used for backfill material within the pay limits in those locations where relative compaction of 95 percent is required. Such sand will normally have a specified gradation and sand equivalent. The sand may be available on the job site, but in many cases it must be ordered from an outside source, and from 1.7 to 1.9 tons will be required per compacted cubic yard.

The horizontal pay limit for structure backfill is often a vertical plane 1 foot away from a concrete wall. Typically, the true section to be backfilled, will be 1 foot wide at the bottom, bounded by the wall on one side and by a sloping bank on the other.

When imported sand is required in such a case, sufficient sand must be ordered to place a vertical sand section at least 1.5 feet wide against the concrete, since it is not practical to place the sand to an exact line. The imported sand and the material from the excavation are placed in layers, 6 to 8 inches thick and as near as possible to their required locations, by a wheel loader, during the compaction operation.

8.2.1 Unit Costs

Typical unit costs for structure backfill in confined areas are shown in Table 8.2. As used here, a confined area is one that is too small to permit the use of road-type compaction equipment.

Except for the pipe backfill, the compaction is assumed to be done with a small loader-backhoe equipped with a hydraulic backfill compactor. Hand operated backfill compactors are assumed for the pipe backfill, since the backhoe mounted compactors will damage or distort pipes.

The unit costs for labor and equipment include the costs for placing imported sand, when required, as well as the costs for applying the water needed for compaction. The material cost for the imported sand and the cost to supply the water to the site are additional in all cases.

Sandy material may be flooded and compacted with concrete vibrators, when the job conditions permit the excess water to drain without damage to the work. Material was placed in a 6-foot-wide area with a ¾ CY clamshell bucket at an effective rate of 17.5 cubic yards per hour. The material was compacted with 2 two-man crews using one 3HP vibrator per crew. When using concrete vibrators in sand, the vibrator heads must be water-cooled when working. This is achieved by furnishing a constant supply of water to each head with a garden hose.

8.2.2 Mass Structure Backfill

Where the available space permits and where large amounts of structure backfill are involved, mass structure backfill is placed with road-type compacting equipment. In the typical case, however, the job conditions relating to the backfill operation do not permit maximum production to be achieved.

Table 8.2. Unit Costs for Structure Backfill

Type of Structure or Activity (relative compaction)	Labor		Rental		Unit of Measure
	Unit Cost	Index	Unit Cost	Index	
Box culverts (90–95%)	$2.50	$K_L = 13.48$	$3.20	$K_{BH+C} = 31^a$	CY
Box culverts (no testing)	$0.80	$K_L = 13.48$	$1.25	$K_{BH+C} = 31$	CY
Flat slab bridges (95%)	$3.25	$K_L = 13.48$	$4.30	$K_{BH+C} = 31$	CY
Girder bridges (95%)	$2.30	$K_L = 13.48$	$3.60	$K_{BH+C} = 31$	CY
Pipe backfill (90–95%)	$6.80	$K_L = 13.48$	$4.70	$K_{BH+T} = 30^b$	CY

[a] K_{BH+C} = hourly rental rate for a ½ CY wheel loader-backhoe equipped with a hydraulic backfill compactor.

[b] K_{BH+T} = the sum of the hourly rental rates for a ½ CY wheel loader-backhoe and a hand-operated backfill compactor.

Note: The unit costs shown are for work in confined areas, and include the labor cost to apply water. The costs to furnish the water required for compaction are additional. The unit costs should be applied to the true quantity of backfill, which may be greater than the amount shown on the engineer's estimate. On-and-off charges for equipment are additional.

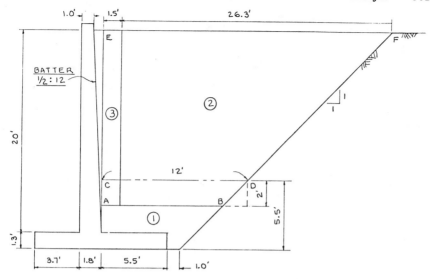

Figure 8.1 Combined structure backfill.

8.2.3   *Combined Structure Backfill*

Maximum economy may require the structure backfill for some structures to be placed by a combination of methods. For example, assume the back face of the retaining wall shown in Figure 8.1 is to be backfilled to line EF. Assume further, that material for the backfill is stockpiled about 100 feet to the right of point *F*, that the wall section shown has an effective length of 200 feet, and that access is available at the ends of the section.

The area to be backfilled is divided into three zones as shown. Zone 1 will be placed with a loader-backhoe equipped with a hydraulic backfill compactor. Zone 2 will be placed with a double 5×5 sheepsfoot roller, and Zone 3 will be placed by hand, concurrently with the material in Zone 2. The hand backfill in Zone 3 is necessary because the sheepsfoot roller cannot work within 1 to 1.5 feet of the wall.

Zone 3 is drawn as shown to simplify the take-off, although this material will, in fact, be placed against the battered back face of the wall.

The lower limit of Zone 2 is determined as follows: line *CD* is located so that its length is 12 feet, which is a minimum width for the sheepsfoot roller. Line *AB* is arbitrarily set 2 feet below line *CD* to increase the area of Zone 2 as much as possible, and the triangular cut at the bank will be made and backfilled at no additional cost during the mass backfill operation.

The backfill is taken off as follows:

Zone 1

$$\text{Area} \quad \frac{10 + 6.5}{2} \times 3.5 - 1.3 \times 5.5 = 21.7 \text{ SF}$$

$$\text{Volume} \quad 21.7 \times \frac{200}{27} \qquad\qquad = 161 \text{ CY}$$

Zones 2 + 3

$$\text{Area} \quad \frac{27.8 + 10}{2} \times 17.8 \qquad = 336.4 \text{ SF}$$

Zone 3

$$\text{Area} \quad 1.5 \times 17.8 \qquad\qquad = 26.7 \text{ SF}$$

$$\text{Volume} \quad 26.7 \times \frac{200}{27} \qquad\qquad = 198 \text{ CY}$$

Zone 2

$$\text{Area} \quad 336.4 - 26.7 \qquad\qquad = 309.7 \text{ SF}$$

$$\text{Volume} \quad 310 \times \frac{200}{27} \qquad\qquad = 2300 \text{ CY}$$

The additional volume of backfill due to the battered back face is $\frac{1}{2} \times 0.8 \times 20 \times 200/27 = 59$ CY, and this volume is added to the volume calculated above for Zone 2. The error that results from neglecting Zone 1 in this procedure is negligible, and the final backfill volumes are taken as follows:

Zone 1   161 CY
Zone 2   2360 CY
Zone 3   198 CY
   Total   2719 CY

Assuming the horizontal pay limits for structure backfill are vertical planes 1 foot outside the neat concrete lines, the engineer's estimate for this portion of backfill would be

$$2719 - \frac{21.3 \times 21.3 \times 200}{2 \times 27} = 1039 \text{ CY.}$$

Since the hand backfill in Zone 3 must be placed concurrently with the mass backfill in Zone 2, it is now necessary to arrive at reasonable, balanced production rates for these two items.

A three-man crew, supplied with material, will compact it, at an effective rate of about 6 CY per hour. This assumes one man operating the compactor, which may be a vibrating plate, a vibrating roller, or a rammer type unit, one man applying water with a garden hose, and one man shoveling and leveling ahead of the compactor.

Thus, two crews will compact the material in Zone 3 in 198/12 = 17 hours while the sheepsfoot will be compacting the material in Zone 2 at an average rate of 2360/17 = 139 compacted CY per hour.

This is a slow production rate for the roller, and we note that it could be doubled if more than two hand backfill crews were used. Considering the cramped work area and the relatively short duration of the backfill operation, however, the 139 CY per hour rate is selected as the fastest practical rate in this particular instance.

The backfill material for Zones 2 and 3 will be pushed over the bank by a bulldozer working at the top, and one 185-hp bulldozer will be adequate.

The unit costs for placing backfill in Zones 1, 2, and 3 are determined as follows:

Zone 1. The backfill in Zone 1 is similar to that required for bridge pier footings, except that the backfill material is further away. The unit costs for girder bridge backfill shown in Table 8.2 will be used, but the rental cost will be increased by an arbitrary $1.00 per CY to cover the cost of moving the material into the hole. The $1.00 per CY allowance assumes that a bulldozer or perhaps a wheel loader will be available somewhere on the job at the time this work is done. Thus the unit costs for the Zone 1 backfill will be:

Labor $2.30 per compacted CY
Rental $4.60 per compacted CY

Zone 2. The material will be furnished by a 185-hp bulldozer, spread by a motor grader, compacted with a 185-hp bulldozer pulling a sheepsfoot roller, and water will be supplied from a water truck. A grading foreman will be required, and his hourly wage rate, including all fringes, is assumed to be $18.50 per hour. The assumed rental rates for the equipment are as follows:

185-hp bulldozer	$ 50 per hour
185-hp bulldozer	50
Double 5 × 5 sheepsfoot roller	5
125-hp blade	39
Water truck	30
Total hourly cost	$174

The unit costs for the Zone 2 backfill will be:

Labor $\dfrac{18.50}{139}$ = $0.14 per compacted CY

Rental $\dfrac{174}{139}$ = $1.25 per compacted CY

Zone 3. Assuming the hourly rental rate for the hand operated backfill compactors to be $5.00 per hour, and that K_L = 13.48, the unit costs for the hand backfill in Zone 1 will be:

Labor $\dfrac{6 \times 13.48 \times 17}{198}$ = $6.94 per compacted CY

Rental $\dfrac{2 \times 5 \times 17}{198}$ = $0.86 per compacted CY

The above unit cost for rental does not include any costs for furnishing the backfill material, since this material will be furnished and spread by the mass backfill equipment.

The time of completion estimate for backfilling the wall is made, using the production rates discussed above together with the production rate for the backfill in Zone 1, which, from Table 8.2, is 31 / 3.60 = 8.6 CY per hour.

The on and off charges for the equipment must be added. Assume the equipment will be moved 4 miles through suburban streets and 11 miles on good country roads. Assume also that the hourly rental rate for a flatbed truck is $30 per hour. Driving time for the flatbed is estimated as 8/25 + 22/40 = 0.9 hour round trip. Allowing 1 hour to load and unload the equipment, the total on and off time per piece is 1.9 × 2 = 3.8^h which is rounded off to 4 hours.

The two bulldozers, the sheepsfoot roller, and the blade will each cost 4 × 30 = $120 to move on and off. The water truck will cost 1 hour's rental, as will the loader-backhoe, which arrives on its own trailer. Thus, the total on and off charges will be:

4 pieces @ $120	$480
Water truck 1 hour	30
Loader-backhoe 1 hour	31
Total	$541

An allowance should be made on the estimate for the cost of the garden hoses used to supply water to the hand backfill operation, and if necessary, the costs for a small (250 gallon) water tank, together with a 2-inch centrifugal pump. It is assumed that the water supply for the water truck is developed under a separate job item.

9

Roadway Excavation and Imported Borrow

The job items for roadway excavation and imported borrow represent similar types of work and, on many projects, the combined costs for these two items will constitute more than 50 percent of the total job cost. Roadway excavation is made with off-highway equipment, wheel tractor-scrapers are commonly used to haul the material, and the haul lengths are dictated by the road profiles.

The type of equipment used to haul imported borrow will depend on whether the hauls are on public roads or off-highway, and on the lengths of hauls required. The haul lengths will depend on the location of the borrow pit with respect to the related embankments; it is imperative that these haul lengths be kept as short as possible.

9.1 Roadway Excavation

Roadway excavation is the excavation required to grade and construct the roadway. The price paid for roadway excavation includes the costs of excavating, sloping, shaping, loading, hauling, spreading, and compacting the material, and the cost of making the grade for the roadway. Roadway excavation is paid for in the cut, and the unit of measure is bank yards.

When the excavated material must be hauled more than 10 stations (1000 feet), a separate item for overhaul may be included in the engineer's estimate.

9.1.1 The Roadway Excavation Take-off

The take-off will include the following items:

1. A check of the engineer's estimated yardage for the item.
2. A take-off of any excavation or work, required by the specifications, which is not included in the engineer's estimate but whose cost is specified to be included in the price bid for roadway excavation.
3. A take-off of that excavation which is neither specified nor included in the engineer's estimate, but which must be done to complete the roadway excavation item.
4. A take-off of the areas on which a grade must be made.
5. A determination of the average haul distances for the various sections of roadway excavation.
6. An estimate of the number of flares, barricades, signs, and flagmen required for traffic control and for the protection of the public during the course of the roadway excavation work.

9.1.1.1 CHECKING THE ENGINEER'S ESTIMATE. The quantities of excavation and embankment are tabulated on the road profiles. On small jobs they are tabulated per station, and on large jobs they may be tabulated for each 10 stations.

The sum of the tabulated excavation quantities should equal the engineer's estimate for the item, but in many cases the estimate will have been arbitrarily increased by about 5 percent, for budgetary reasons.

The tabulated quantities of excavation and embankment will be true quantities. To determine haul distances, one or the other of these quantities must be adjusted by the grading factor.

9.1.1.2 SPECIFIED WORK NOT INCLUDED IN THE ENGINEER'S ESTIMATE. The specifications for roadway excavation frequently require performing a number of related work items whose costs are specified as included in the price bid for roadway excavation. Such work may include certain clearing and grubbing operations, saw cutting of pavements, compacting original ground, undercutting and recompacting material in certain areas, or the benching of hillsides against which embankments will be constructed. The additional excavation required because of shrinkage resulting from such compacting or recompacting work will not be included in the excavation quantities tabulated on the profiles.

Because of the inherent imprecision in determining the grading factor, and because of the additional compaction discussed above, some roadway excavation jobs will end up short of dirt. In such cases the costs of obtaining and placing the additional material required to complete the embankments are normally covered under a contract change order.

When embankments are constructed on hillsides or against existing embankments, the existing slopes must be cut unto, horizontally, for a specified distance, and the work is brought up in layers. The material thus cut is recompacted with the embankment material as the work progresses, and the excavation required is not included in the engineer's estimate.

The take-off for such bench excavation is based on assumed typical sections, the excavation is made with a bulldozer, and no other costs need be considered.

9.1.1.3 ADDITIONAL EXCAVATION REQUIRED FOR CONSTRUCTION.   A typical cross section for a freeway ramp constructed of embankment is shown in Figure 9.1, together with a sketch used to determine the additional excavation required to construct the shoulders.

The hinge points, designated *HP,* are located at the intersections of the upper surface of the roadway and the side slopes. The area between the hinge points is the roadbed, and area *ABCD* is the structural section, which, in this case, consists of 0.5 feet of asphaltic concrete on 1.05 feet of aggregate base.

The embankment section must be completed before placing materials in the structural section. To complete the embankment, a grade must be made on the shoulders and on the subgrade, which is line *BC* on the cross section.

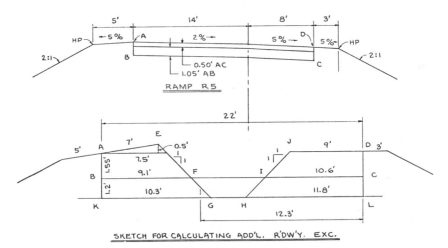

Figure 9.1 Road cross section.

Since the 5-foot shoulder on the left and the 3-foot shoulder on the right are both too narrow to permit road-building equipment to operate, enough additional embankment to furnish a minimum total top width of 12 feet, must be placed at each shoulder. The additional excavation required to place and remove this temporary embankment material will not be included in the engineer's estimate, and thus no payment will be made for doing this work.

Referring to the sketch on Figure 9.1, areas *ABFE* and *CDJI* represent additional embankment required above the subgrade.

$$\text{Area } ABFE = 14.7 \text{ SF}$$
$$\text{Area } CDJI = 15.2 \text{ SF}$$

The triangular area at the top of *ABFE* results from the 7 percent difference in the slopes between the shoulder and the structural section. A similar situation near point *J* is neglected, since the difference in slopes is only 3 percent and the length involved is only 1 foot.

Assuming the length of the ramp to be 700 feet, the total volume of additional material above the subgrade is

$$700 \times \frac{14.7 + 15.2}{27} = 775 \text{ CY}$$

After the shoulders have been completed, the additional material must be removed and placed in an embankment. On wide structural sections where the distance *FI* is about 14 feet or more, the subgrade is left low enough, in this area, to receive the additional material, which is excavated and moved to its final position with a motor grader. When structural sections are too narrow to permit the above procedure, the additional material is excavated, hauled, and placed into a nearby embankment.

In both procedures noted above, one additional roadway excavation operation is required for the material used to temporarily widen the shoulder sections, which, in this case, amounts to 775 CY. The cost to excavate this material is based on the unit costs for roadway excavation made within the free-haul limit.

Sometimes the structural sections are narrow, there are no nearby embankments, and the additional material must be placed below the subgrade of the structural section in which it is used. This procedure requires additional unpaid-for excavation, and it will be necessary to widen the shoulder sections one side at a time.

In such cases, the embankment section is constructed to line *KL,* which is parallel to, and below, the subgrade. Since area *CDJI* is greater than area *ABFE,* the right shoulder is completed first, which requires area *DJHL* to be placed.

Upon completion of the right shoulder, area *DJHL* is removed and placed in the left shoulder area *AKGE;* the excess material is used in the shoulder section to the left of line *AK*.

After completion of the left shoulder, area *ABFE* is removed and placed in area *CFGL,* which requires that the distance *BK* be selected so that areas *ABFE* and *CFGL* are equal. Assuming distance *BK* to be about 1 foot, the average width of area $CFGL = 22 - 9.1 - 0.5 = 12.4$ feet, and we have

$$BK \ (12.4) = 14.7$$

or

$$BK = 1.2 \text{ feet.}$$

The areas for the shoulder sections shown in Figure 9.1 are calculated and listed as follows:

Area *DJHL* $= 28.6$ SF
Effective area $DJHL = 28.6/0.85 = 33.6$ SF
Area *AKGE* $= 26.4$ SF
Area *CBKL* $= 26.4$ SF

Effective area $CBKL = 26.4/0.85 = 31.1$ SF

An assumed grading factor is applied to areas *DJHL* and *CBKL* since both will, in effect, be filled with material from the cut. The additional, unpaid excavation is determined as follows:

+33.6 SF	Excavate and place in right shoulder
+28.6 SF	Excavate from right shoulder and place in left shoulder
−2.2 SF	Area *DJHL* − Area *AKGE*. This represents excess material placed to the left of line *AK*. This material is paid for
+14.7 SF	Excavate area *ABFE* and place in area *CFGL*
−31.1 SF	Effective area *CBKL* is paid for under the roadway excavation item
43.6 SF	Additional, unpaid excavation area

Assuming the ramp to be 700 feet long, the additional unpaid excavation is

$$43.6 \times \frac{700}{27} = 1130 \text{ CY.}$$

To save time, the estimator calculates this yardage by more approximate methods. The triangular area at the top will be neglected, and the remaining areas are assumed to be rectangular in shape. Since typical jobs may have 20 to 40 ramps, a significant amount of additional excavation may be involved, and its additional cost should not be overlooked.

9.1.1.4   AREA TAKE-OFF FOR MAKING GRADE.   The final operation for the roadway excavation item is making the grade for the roadbed. To determine the required areas, widths are taken from the typical road cross sections, and are the total widths between hinge points. The corresponding lengths are taken from the road plans, and the resulting areas are expressed in square yards.

Referring to Figure 9.1 and assuming that R5 is 700 feet long, the area for making the grade on this ramp would be $30 \times 700 \times \frac{1}{9} = 2330$ SY. When the unit costs for making the grade on roadbeds are applied to the total of all such areas, the resulting costs will include the costs for making the grade on the shoulders as well as on the subgrade of the structural sections.

It is not necessary to take off the areas of the cut or embankment slopes. The allowable grade tolerance is normally ± 0.5 foot for cut slopes, and from ± 0.5 to ± 1.0 foot for embankment slopes. With proper care, these tolerances can be met fairly well as the work is done, but the embankment slopes may require trimming, under the item for finishing roadway.

The median is that portion of a divided highway separating the traveled ways, and includes the inside shoulders. In taking off the median areas for grade making purposes, however, the widths are taken between the inside shoulders. The median areas are kept separate from the roadbed areas, since different unit costs apply.

Many jobs have items for both roadway excavation and imported borrow. In such cases the grade-making areas for the entire job are taken off in one operation, and no attempt is made to separate the embankment sections made under either item. The total areas so determined may be somewhat arbitrarily apportioned between the two items, or the entire grade making operation may be charged to the larger item.

9.1.1.5   AVERAGE HAUL DISTANCES.   The determination of the average haul distances relating to the different segments of excavation is an important part of the roadway excavation take-off, since haul lengths directly affect the cycle times for the hauling equipment.

The determination of haul distances requires that certain cut volumes be balanced with certain fill volumes. To achieve such an earthwork balance, either the cut or the fill volumes, tabulated on the road profiles, must be adjusted by the grading factor.

9.1.2 Haul Distances by Arithmetic Methods

On small jobs, haul distances may be determined by inspection or arithmetically, as shown in Table 9.1. This table assumes that excavation and embankment quantities are tabulated on the road profiles in two-station intervals, and that the grading factor is 0.85.

The quantities in columns 2 and 4 are taken directly from an assumed road profile, the embankment quantities being considered negative, and the excavation quantities being considered positive. The grading factor is applied to the embankment quantities, and the adjusted embankment quantities are listed in column 3. Column 5 contains the net cubic yards per interval, and is the algebraic sum of the quantities in columns 3 and 4. The center of mass (actually, the center of volume) for each quantity, listed in column 5, is assumed to be at the center of each two-station interval.

The net volumes in column 5 are balanced in column 6, starting at the grade point, station 11, where the job passes from cut to fill. The 53 CY of fill required at station 11 is made from the excavation at station 9. This leaves 747 CY of excess excavation at station 9, which is taken to station 13. This leaves 194 CY of fill still required at station 13, which is taken from the excavation at station 7. This process is continued, and finally, at station 21, we have an excess of 24 CY of cut, which checks the column 5 total.

The haul distances for the excavation quantities are taken from column 6, where each of the excavation quantities appears both at its station of origin and at the station to which it is hauled. The distances and quantities are tabulated as follows:

Stations		Haul Distance	
From	To	(Stations)	CY
9	11	2	53
9	13	4	747
7	13	6	194
7	15	8	906
5	15	10	271
5	17	12	929
3	17	14	12
3	19	16	988
1	19	18	165
19	21	2	141
		Total CY hauled	4406

Table 9.1. Haul Distances—Arithmetic Method

(1)	(2)	(3)	(4)	(5)	(6)
Station	Embankment CY	Embankment/0.85 CY	Excavation CY	Net CY	Earthwork balance
0	− 200	− 235	+ 400	+ 165	−165 = 0
2	—	—	+1000	+1000	− 12 = +988
4	—	—	+1200	+1200	−271 = +929
6	—	—	+1100	+1100	−194 = +906
8	—	—	+ 800	+ 800	− 53 = +747
10	− 300	− 353	+ 300	− 53	+ 53 = 0
12	− 800	− 941	—	− 941	+747 = −194
14	−1000	−1177	—	−1177	+906 = −271
16	− 800	− 941	—	− 941	+929 = − 12
18	−1100	−1294	—	−1294	+988 + 165 = −141
20	− 200	− 235	+ 400	+ 165	−141 = + 24
22					
Total	−4400	−5176	+5200	+ 24	

313

The total CY hauled, above, does not equal the total of the cubic yards of excavation tabulated on the profile and listed in column 4 of Table 9.1. This results from the use of net quantities for those segments consisting of both cut and fill. Referring to Table 9.1, we have 235 CY of excavation at station 1, 300 CY at station 11, and 235 CY at station 21, or a total of 770 CY of excavation, which is not included in the haul yardage. As discussed in the following section, this is called sidehill excavation, and it, together with the 24 CY of waste, will appear on the engineer's estimate, and will be paid for. The sum of the haul yardage, the sidehill excavation, and the waste equals 5200 CY, which checks the total for column 4, Table 9.1.

The average haul distance from the center of mass of the excavation to the center of mass of the embankment for the material tabulated above is $(2 \times 53 + 4 \times 747 + 6 \times 194 + 8 \times 906 + 10 \times 271 + 12 \times 929 + 14 \times 12 + 16 \times 988 + 18 \times 165 + 2 \times 141) \div 4406 = 10$ stations. In addition, the 794 CY of sidehill excavation and waste will be moved about one station.

For estimating purposes, however, the above haul yardage would be broken down into yards hauled 10 stations or less, and yards hauled more than 10 stations. For these purposes, the preceding tabulation is re-stated as follows:

Haul Distance (stations)	CY
2	53
4	747
6	194
8	906
10	271
2	141
Total yards (hauled 10 stations or less)	2312

and

Haul Distance (stations)	CY
12	929
14	12
16	988
18	165
Total yards (hauled more than 10 stations)	2094

Calculating the average haul distances for these yardages, as before, we get

2312 CY hauled 6 stations

and

2094 CY hauled 14 stations.

In addition, the 794 CY of sidehill excavation and waste will be moved about one station.

The 10-station average haul for the entire section can also be determined without resorting to the arithmetical balancing procedure shown in Table 9.1, column 6. To do this, the centers of mass must be located for the section in cut and for the section in fill. Taking the data from Table 9.1, we have

Station	Net CY Excavation
1	165
3	1000
5	1200
7	1100
9	800
Total	4265 CY

and

Station	Net CY Embankment
11	53
13	941
15	1177
17	941
19	1294
Total	4406 CY

Taking moments about station 0, the station of the center of mass for the excavation section is: $(1 \times 165 + 3 \times 1000 + 5 \times 1200 + 7 \times 1100 + 9 \times 800) \div 4265 = 5.6$. Again, taking moments about station 0, the station of the center of mass of the embankment section is found to be station 16.1.

Thus we have

4265 CY haul $(16.1 - 5.6) = 10.5$ stations

and

$$\frac{141}{4406} \text{ CY haul } (21 - 16.1) = 4.9 \text{ stations}$$
$$4406 \text{ CY total}$$

The average haul for this yardage is $(4265 \times 10.5 + 141 \times 4.9) \div 4406$ = 10 stations. As before, the 794 CY of sidehill excavation and waste will be moved about one station.

9.1.3 Haul Distances from Mass Diagrams

On large jobs, haul distances are determined by use of mass diagrams which are often roughly plotted on quadrille paper by the estimator. A mass diagram is a curve, plotted in connection with a road survey having excavation and embankment quantities. As in the preceding example, cut quantities are considered positive and fill quantities are considered negative.

The ordinates for the mass diagram are calculated from the excavation and the embankment quantities as described below. Since the mass diagram will be used to determine earthwork balances, either the cut or the fill volumes, from the road profile, must be adjusted by the grading factor prior to being used to calculate the ordinates.

When the grading factor is constant for all of the cut sections, either the cut volumes or the fill volumes may be adjusted. In such cases, however, it is advantageous to adjust the embankment volumes, since the mass diagram ordinates will then represent true cut volumes (pay units) and the volumes of cut necessary to make the fills.

When the cut material consists of different types having different grading factors, the cut volumes must be adjusted. In such cases the mass diagram ordinates will represent cut volumes available for fill, and the true fill volumes.

In the following examples we will assume the grading factor is constant for the entire section considered, and will adjust the embankment quantities, dividing each of them by the grading factor. Thus, throughout this discussion of mass diagrams, the terms "excavation" or "cut" will always refer to true excavation volumes, and the terms "embankment" or "fill" will always refer to the adjusted quantities.

Typical mass diagrams are shown in Figures 9.2 and 9.3. The abscissas, (horizontal distances) on a mass diagram are the roadway stations. The ordinate (vertical distance) at any station is the algebraic sum of the net cut and fill yardages up to that station. Thus, at station 0 the ordinate will be zero, at station 1 the ordinate will be the algebraic sum of the cut and

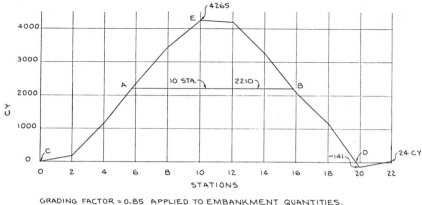

GRADING FACTOR = 0.85 APPLIED TO EMBANKMENT QUANTITIES.

SCALE: HORIZ. 1" = 2 STA.
VERT. 1" = 1000 CY

Figure 9.2 Mass diagram.

fill volumes between station 0 and station 1, and at station 2 the ordinate will be the algebraic sum of the cut and fill volumes between station 0 and station 2.

If, on a given profile, there is sidehill work or a grade point between, say, stations 8 and 9, a cut volume and a fill volume will both be tabulated on the road profile, for that particular interval. Since net volumes (the algebraic sum of the cut and the fill volumes) are used to calculate mass diagram ordinates, the cut volume necessary to balance the adjusted fill volume for that particular interval will not appear on the mass diagram, even though the cut must be made, will appear on the engineer's estimate, and will be paid for.

On jobs where mass diagrams will be used, therefore, a separate total for such sidehill excavation must be made at the time the engineer's estimate for the roadway excavation is being checked. For any interval, the sidehill excavation will be numerically equal to the smaller of the cut or the adjusted fill volumes for that interval. As a check, the sum of the total sidehill excavation and the total excavation from the mass diagram should equal the total excavation tabulated on the profile.

9.1.3.1 PROPERTIES OF THE MASS DIAGRAM.   The essential properties of the mass diagram are:

1. Maximum or minimum points on the curve occur at grade points, where the profile changes from cut to fill or from fill to cut.

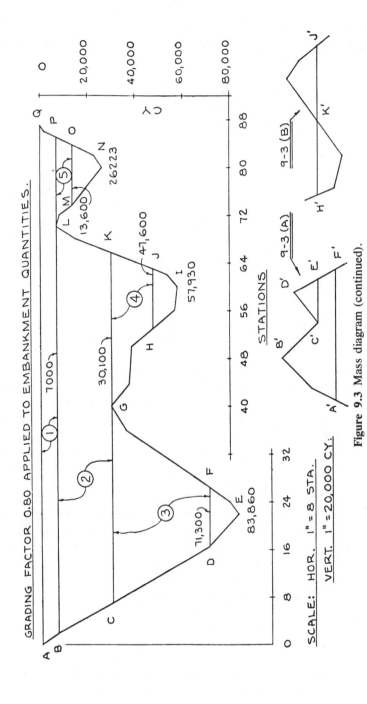

Figure 9.3 Mass diagram (continued).

2. An upward-sloping curve indicates excavation, and a downward-sloping curve indicates fill. The slopes of the curve vary with the volume of cut or fill per linear foot along the road centerline.
3. A horizontal line intersecting the curve at any two points indicates a balance of cut and fill volumes between the two points of intersection, and the quantities intersepted by any two horizontal lines will balance.
4. Since the abscissas are stations and the ordinates are cubic yards, the areas on the curve represent haul, where haul is defined as volume × distance, and is measured in station yards.

9.1.3.2 HAUL DISTANCES. Properties 3 and 4, above, enable haul distances to be determined. Referring to Figure 9.2, excavation *AE* will balance fill *EB* (property 3). Since area *AEB* is expressed in station yards (property 4), the haul distance between the centers of volume of excavation *AE* and fill *EB* is determined by dividing area *AEB* by the ordinate distance from line *AB* to the maximum point *E*. The haul distances for the example given in Table 9.1 will now be determined by the use of a mass diagram. The mass diagram ordinates are calculated in Table 9.2, and the mass diagram is shown in Figure 9.2.

The quantities shown in the embankment and excavation columns of Table 9.2 are assumed to be taken directly from the tabulated quantities on the road profile. The grading factor is assumed to be 0.85, and it is assumed the quantities are tabulated on the profiles in two-station intervals.

Balancing line *AB* is positioned on the mass curve so as to be exactly 10 stations in length, for reasons discussed below in section 9.1.3.3. The excavation from *A* to *E* will exactly balance the embankment from *E* to *B*, and the excavation from *C* to *A* will exactly balance the embankment from *B* to *D*.

The scale of the mass diagram is two stations per inch horizontally and 1000 CY per inch vertically, which equals 2000 station yards per square inch. The area under the curve and above line *AB* is measured with a planimeter or by scaling, and is 6.38 square inches or 12760 station yards. The ordinate to line *AB* is scaled, and equals 2210 CY. The ordinate from line *AB* to the maximum point *E* is $4265 - 2210 = 2055$ CY. The haul distance between the centers of volume of excavation *AE* and embankment *EB* is $12760/2055 = 6$ stations.

Similarly, the area between lines *AB* and *CD* measures 15.66 square inches or 31320 station yards, and the haul distance between the centers of volume of excavation *CA* and embankment *BD* is $31320/2210 = 14$ stations.

Table 9.2. Mass Diagram Ordinates

Station	Embankment CY	Embankment 0.85	Excavation CY	Net Volume CY	Mass Diagram Ordinate	Sidehill Excavation CY
0					0	
2	− 200	− 235	+ 400	+ 165	+ 165	235
4			+ 1000	+ 1000	+ 1165	
6			+ 1200	+ 1200	+ 2365	
8			+ 1100	+ 1100	+ 3465	
10	− 300	− 353	+ 800	+ 800	+ 4265	
12	− 800	− 941	+ 300	− 53	+ 4212	300
14	− 1000	− 1177		− 941	+ 3271	
16	− 800	− 941		− 1177	+ 2094	
18	− 1100	− 1294		− 941	+ 1153	
20	− 200	− 235		− 1294	− 141	
22			+ 400	+ 165	+ 24	235
Total	− 4400	− 5176	+ 5200	+ 24	—	770

By addition, the area under the curve and above line *CD* = 44080 station yards, and the haul distance between the centers of volume of excavation *CE* and embankment *ED* = 44080/4265 = 10 stations.

Referring to the mass diagram, we note that the 141 CY of embankment to the left of station 20 is balanced by the 141 CY of excavation to the right, that the haul for this material is about one station, and that 24 CY of material will be excavated and wasted.

The sidehill excavation from Table 9.2 is 770 CY, and a summary of the pertinent information for the section of roadway being considered is:

> 2055 CY haul 6 stations
> 2210 CY haul 14 stations
> $\underline{935}$ CY haul 1 station ± (770 + 141 + 24 = 935)
> $\overline{5200}$ CY Total.

9.1.3.3 OVERHAUL. On some jobs there is an item for roadway excavation and an item for overhaul. In such cases, the roadway excavation item, as set forth in the specifications, will include the costs of excavating, placing, and hauling the material up to the limit of free haul, which is normally 10 stations. The overhaul item then includes the costs of hauling the material beyond the free haul limit, and the unit of measure for the item is usually station yards.

Mass diagrams are always used to determine the amount of the overhaul item. In the preceding example, the balancing line *AB* was set at the limit of free haul, which is 10 stations. The average haul distance for excavation *CA* when moved to embankment *BD* was 14 stations, the overhaul distance is (14 − 10) = 4 stations, and the total overhaul for the 22 station section is 2210 CY × 4 stations = 8840 station yards.

Figure 9.3 illustrates a typical mass diagram for a section of freeway. Lines *DF, HJ,* and *MO* are 10 stations long, which is the free haul distance. Thus, excavations *EF, IJ,* and *NO* balance fills *DE, HI* and *MN,* and none of this material is moved more than 10 stations. The ordinates to the free haul limit lines are scaled and written on the diagram as shown.

Line *AQ* is the horizontal axis for the diagram, and lines *BP* and *CK* are balancing lines drawn to determine haul distances. Since these lines are drawn through the maximum points *G* and *L,* the ordinates to them are taken from the table of ordinates calculated for the diagram. As before, cuts *QP, PO, LK, KJ,* and *GF* balance fills *AB, LM, BC, GH,* and *CD.*

Table 9.3 shows the calculations required to determine the overhaul distances. The ordinate distances for each of the five sections are listed in column 2, and the areas of the sections are listed in column 3.

Table 9.3.   Overhaul

(1)	(2)	(3)	(4)	(5)	(6)	(7)
Sec. No.	Ordinate CY	Area (sq in.)	SYH	SH	SOH	SYOH
1	7000	3.709	593,400	84.77	74.77	523,400
2	23100	9.051	1,448,000	62.68	52.68	1,217,000
3	41200	5.270	843,200	20.47	10.47	431,400
4	17500	1.878	300,500	17.17	7.17	125,500
5	6600	0.492	78,700	11.92	1.92	12,700
					Total	2,310,000

Station yards of haul are listed in column 4, and these quantities are obtained by multiplying the areas in column 3 by 160000 station yards per square inch, which is the scale of the diagram. The stations of haul listed in column 5 are obtained by dividing the station yards of haul by their respective ordinate distances.

The stations of overhaul listed in column 6 are the stations of haul less the free haul distance, which is 10 stations. Since the grading factor, as noted on the diagram, was applied to the embankment quantities, the ordinates listed in column 2 represent true cut yardages. The station yards of overhaul listed in column 7 are obtained by multiplying these ordinates by their respective stations of overhaul.

As discussed previously, the grading factor must be applied to the cut volumes when the grading factor varies for the different sections of cut. In such a case, the ordinates listed in column 2 would be reduced by the grading factor and would not represent true cut yardages. The areas plotted from such ordinates would also be reduced by the same amount, but the values calculated for stations of haul and for stations of overhaul would not be affected. To determine the station yards of overhaul in such a case, however, the adjusted ordinate is divided by the grading factor to obtain the true cut volume, which is then multiplied by the stations of overhaul.

Two special cases which often occur in connection with mass diagrams are also illustrated in Figure 9.3. Figure 9.3(A) shows a double loop within the free haul limit, represented by line $A'F'$. In such a case, the total yardage of free haul is indicated by the sum of the ordinate from line $A'F'$ to B' and the ordinate from line $C'E'$ to D'.

Figure 9.3(B) represents a pair of loops, and $H'J'$ is a balancing line where both $H'K'$ and $K'J'$ exceed the free haul distance. In such a case, minimum overhaul between points H' and J' will occur when the balancing line is positioned so that $H'K'$ and $K'J'$ are equal.

9.1.4 Unit Costs for Roadway Excavation

Many earthwork contractors bid roadway excavation strictly on an overall comparison basis. The job under consideration is compared with past, completed work of a similar nature, regarding quantities, type of material, and haul lengths, and the overall unit costs realized on the past work are adjusted, for comparison, to reflect the wage and rental rates applicable to the new job.

An overall unit cost for the job being bid is then selected, based on the contractor's judgment, the adjusted overall unit costs from the past jobs, and a review of the recent bid prices for similar work as reported in the daily construction services. In most cases, such contractors must wait until the roadway excavation is completed to find out whether money was made or lost on the item.

Although many such contractors manage to stay in business, all successful contractors know of instances where large amounts of money were lost, or where bankruptcy occurred, because the job bid was not analyzed in sufficient detail to reveal major and costly differences between it and the work with which it was compared.

A far better procedure requires that the roadway excavation item be broken down into similar segments of work, that each of these segments be analyzed, and that a unit cost be derived for each segment, based on estimated production rates for the equipment that will be used. These derived unit costs are then compared with the adjusted unit costs realized on past work for similar operations; where appreciable differences are observed, the derived unit costs are reviewed and revised or justified.

The contractor who follows this procedure will, if awarded the job, know exactly what production rates must be achieved for each segment, can intelligently monitor and control the work during construction, and will sleep better at night.

Unlike, say, the structure concrete item, it is not possible to precisely determine the final, on-completion amount of roadway excavation in advance. This results from the inherent imprecision in the selection of the grading factor and in the method used to calculate the volumes, from additional compaction, subsidence, and from the permitted tolerances in grades and in the location of side slopes.

As discussed in Section 1.6.2.2, and as illustrated in Chapter 4, it is not prudent to mark up the unit cost for an item that may under-run the quantity shown on the engineer's estimate. Therefore, the roadway excavation item is normally bid at, or very close to, the contractor's bare unit cost for doing the work, and many contractors bid the item at their bare unit cost for excavating, loading, hauling, and compacting the material. In such cases, the fixed costs for moving equipment on and off, for making the grade (which depends on areas which can be precisely determined), and for definite quantities of included work such as curb or pavement removal will appear, together with the markup for the item, in the price bid for one or more of the other items. The items so marked up will be lump sum items or items for which the engineer's estimate can be checked during the take-off. Such items should be early items, and the item for clearing and grubbing, which is normally bid lump sum, is frequently used to carry the markup for items whose final quantities are indeterminate or which will under-run the engineer's estimate.

9.1.5   Wheel Tractor-Scraper Excavation

Different makes and models of wheel tractor-scrapers operate fully loaded, under typical job conditions, at about the same speeds. The top speed, loaded, is about 30 miles per hour for both standard and elevating scrapers, except that the smaller sizes of elevating scrapers have a top speed of about 22 miles per hour. The average travel speed on a given job

Figure 9A Caterpillar Wheel Tractor-Scraper 28 CY struck, 38 CY heaped.

depends on the condition of the haul road, its percent of grade, and on the length of haul.

The overall costs for scraper excavation are greatly affected by the quantities involved. On small jobs, the fixed costs are high with respect to the volume of material to be excavated, and the work will normally be completed before a high-production string of equipment could be organized.

When the quantities are large, the cost estimate should always be based on the contractor's own cost and production records from past jobs. The various factors affecting wheel-scraper cycle times are discussed below, and empirical formulas are given for the times required to complete the components of the cycle times. Although these formulas fit the production data from a number of jobs having average conditions and average operators working under first-class supervision, the data used by different contractors for calculating scraper cycle-times vary. The formulas are presented primarily to illustrate the factors to be considered in developing production records, and in estimating scraper excavation costs. These formulas should not be blindly used by inexperienced estimators.

The cost to excavate and haul with rubber tired scrapers depends upon their cycle times.

$$T = L + H + U \tag{9.1}$$

where

T = cycle time in minutes.
L = loading time in minutes.
H = hauling time in minutes.
U = unloading time in minutes.

The loading time will depend on the size of the scraper, and on the number and the size of the push-tractors used. Based on a 60-minute hour and assuming average working conditions, the loading times will average about as follows:

$$L_1 = \frac{V}{15} \text{ but not less than } 1.0^m \tag{9.2}$$

or

$$L_2 = \frac{V}{25} \text{ but not less than } 0.8^m \tag{9.3}$$

where

L_1 = loading time in minutes when one 270 HP push-tractor is used.
L_2 = loading time in minutes when two 270 HP push-tractors or one 385 HP push-tractor are used.
V = number of bank yards per load.

Two 385 HP-plus push-tractors are often used with scrapers having heaped capacities greater than about 30 CY, and in such cases formula (9.3) applies. When elevating (self-loading) scrapers are used, the loading time, based on a 60-minute hour, will be about 1.0 minute, and no push-tractors are required.

Based on a 60-minute hour, average working conditions, and well-maintained haul roads that are essentially level with no prolonged grades, average travel speeds for wheel tractor-scrapers may be roughly estimated by the following formula.

$$S = 445 + 0.114 \, D_R \qquad (9.4)$$

where

D_R = round trip haul distance in feet.
S = haul speed in feet per minute for values of D_R ranging from 2000 to 14000 feet.

When D_R is less than 2000 feet, the scrapers will cycle at an effective rate of 11 cycles per hour, as discussed in Section 9.1.6.

When the loaded scrapers must travel up a grade, the average round-trip travel speeds will be about

$$S_{4G} = 1100 + 0.01 \, D_R \qquad (9.4.1)$$

$$S_{6G} = 980 + 0.01 \, D_R \qquad (9.4.2)$$

$$S_{8G} = 750 + 0.01 \, D_R \qquad (9.4.3)$$

where

S_{4G} = haul speed in feet per minute (4 percent grade) for values of D_R ranging from 7000 to 14000 feet.
S_{6G} = haul speed in feet per minute (6 percent grade) for values of D_R ranging from 5000 to 14000 feet.
S_{8G} = haul speed in feet per minute (8 percent grade) for values of D_R ranging from 3000 to 14000 feet.

When D_R, on the grades, is less than the lower limits given above, formula (9.4) applies.

In all cases, the average hauling time, in minutes, is

$$H = \frac{D_R}{S} \qquad (9.5)$$

Based on a 60-minute hour, the unloading time will be about

$$U = \frac{V}{90} \text{ but not less than 0.2 minutes} \qquad (9.6)$$

where

$U =$ unloading time in minutes.
$V =$ number of bank yards per load.

The number of scrapers used on a given haul cannot exceed the production rate for the push-tractors used. The number of scrapers that can be loaded by one push-tractor or by one pair of push-tractors is

$$N = \frac{T}{P} \qquad (9.7)$$

where

$N =$ number of scrapers.
$T =$ cycle time for scrapers in minutes.
$P =$ cycle time for pushers in minutes.

The cycle time for the pushers will be about

$$P_1 = \frac{V}{12} \text{ but not less than 1.5 minutes} \qquad (9.8)$$

or

$$P_2 = \frac{V}{18} \text{ but not less than 1.1 minutes} \qquad (9.9)$$

where

P_1 = pusher cycle time when one 270 HP tractor is used.
P_2 = pusher cycle time when two 270 HP tractors or one 385 HP tractor
are used.
V = number of bank yards per load.

Production rates, determined by the above formulas, will be based on a 60-minute hour. As illustrated in the examples which follow, these production rates are adjusted by a 45- or a 50-minute hour when deriving the unit costs to be used on the estimate sheet.

9.1.6   Example: Roadway Excavation—Small Jobs

The roadway excavation quantities shown in Table 9.2 and Figure 9.2 are typical for the roadwork required when a flat slab bridge or a box culvert is constructed. The quantities are summarized again as follows:

	2055 CY haul 6 stations
	+ 935 CY haul 1 station
or	2990 CY haul less than 10 stations
and	2210 CY haul 14 stations

The hourly equipment rental and wage rates are assumed as follows:

Scraper 10/12.5 CY struck/heaped	$48
Bulldozer 270 HP	$60
5×5 sheepsfoot roller	$ 7
Blade 125 HP	$39
Water truck 3M gal.	$30
Grade foreman	$18.50
Grade checker	$16.10

In practice, the applicable rental and wage rates are written at the top of the worksheet used to determine unit costs, and this worksheet is kept immediately behind the estimate sheet for the roadway excavation item.

The cycle time for work within the free haul limit is slow. The area is congested, since the compaction equipment is working close to the cut, and the distances are too short to permit high speeds. On both large and small jobs, the effective production rate for wheel tractor-scrapers working within the free haul limit will be about 11 cycles per hour.

Assuming the material to be a sandy loam, and that one 270 HP pusher will be used, we conclude that the number of bank yards carried per load will be about $(10 + 12.5) \times \frac{1}{2} = 11.2$ BCY, which is equivalent to a load factor of 0.90 applied to the heaped capacity of the scraper. Thus, one scraper will excavate and haul $11 \times 11.2 = 123$ BCY/hr. or 984 BCY per 8-hour day. Considering the quantities involved, two scrapers will be used.

One grade foreman will be required, together with two grade checkers: one at the cut, and one at the fill. Since the production rate is low, and since the fill is close to the cut, the haul roads will be maintained by the blade and the water truck used on the fill at no additional cost.

The unit costs for excavating the material within the free haul limit are determined as follows:

Hourly rental cost:

Push tractor 270 HP	$60
2 scrapers @ $48	96
Blade 125 HP	39
Tractor plus sheepsfoot	67
Water truck	30
Total	$292

Unit rental cost $\dfrac{\$292}{2 \times 123} = \$1.19$ per CY

Unit labor cost $\dfrac{18.50 + (16.10 \times 2)}{2 \times 123} = \$0.21$ per CY

To estimate the unit costs for the work outside the free haul limit, the cycle time must be determined. From equation (9.2) we have

$$L_1 = \frac{11.2}{15} = 0.75, \text{ minimum } = 1.0 \text{ minutes.}$$

From equations (9.4) and (9.5), the travel time for a round trip haul of 28 stations is

$$S = 445 + 0.114 \times 2800 = 764 \text{ ft/min.}$$

and

$$H = \frac{2800}{764} = 3.7^{m}$$

This haul time assumes that the haul road is essentially level. From equation (7.6), the unloading time is

$$U = \frac{11.2}{90} = 0.12, \text{ minimum} = 0.2^m$$

and the total cycle time is $1.0 + 3.7 + 0.2 = 4.9^m$.

The scrapers carry 11.2 bank yards per load, and the effective production rate, based on a 50-minute hour, is

$$\frac{50}{4.9} \times 11.2 = 114 \text{ CY per hour scraper.}$$

Two scrapers will be used, and the unit costs for equipment and for labor are:

$$\text{Equipment } \frac{292}{2 \times 114} = \$1.28 \text{ per CY}$$

$$\text{Labor } \frac{18.50 + (2 \times 16.10)}{2 \times 114} = \$0.22 \text{ per CY}$$

As before, the haul roads will be maintained by the blade and the water truck used on the fill at no additional cost. In both instances we assume that the water supply is on the job, and the costs to develop the water supply are covered under a separate job item. In this example only two scrapers are used, the production rate is low, and it is not necessary to check the cycle time for the push-tractor nor the production rates for the compaction and watering equipment.

9.1.6.1 UNIT COST FOR OVERHAUL. As illustrated in the preceding example, the costs to excavate and move dirt beyond the free haul distance are based on the actual hauls involved, and a unit cost for overhaul is not considered. In the example, the scrapers are traveling at an average speed of 764 feet per minute = 382 stations per hour, based on a 50-minute hour. The unit cost for overhaul would be

$$C = \frac{2R}{VS}$$

where

C = unit rental cost for overhaul in dollars per station yard.
R = hourly rental rate for the hauling unit.
V = bank yards carried per load.
S = average haul speed in stations per hour.

The factor 2 is required in the numerator since the overhaul item is based on one-way haul distances, while the hauling units must make round trips. In our example, the unit rental cost for overhaul, based on a 50-minute hour, would be:

$$\frac{48 \times 2}{11.2 \times 382} = \$0.0224 \text{ per station yard.}$$

This calculation does not reflect the cost for maintaining haul roads in the overhaul area, and the total cost for such maintenance is included in the estimated cost for the roadway excavation.

Since the estimated costs for roadway excavation include the costs for overhaul, no cost for overhaul is shown in the cost column of the estimate summary sheet. The unit price bid for overhaul is selected somewhat arbitrarily. If the pay quantity for the overhaul item is based on field calculations by the resident engineer, the pay quantity may under-run the engineer's estimate, and the item would be bid at a unit price below the theoretical unit cost. If the pay quantity for the item is specified to be the quantity shown on the engineer's estimate, the item is, in effect, a lump-sum item and would be bid at a unit price above the theoretical unit cost. The estimated cost for the overhaul item is deducted from the estimated cost for the roadway excavation, which includes overhaul, to determine the bare unit cost of the roadway excavation item alone.

As illustrated in the sample estimates in Chapter 4, the difference between the balanced unit price and the unit price actually bid for any item is automatically absorbed by the unit prices bid for some of the other items, since the totals for the cost columns and the price bid columns of the estimate summary sheet must balance.

9.1.7 Example: Roadway Excavation—Large Jobs

On jobs having large quantities of roadway excavation, higher production rates and greater efficiencies can be achieved. The procedures for estimating the costs of such work are similar to those illustrated in the preceding example, except that care must be taken to balance the pushing, hauling, and compaction operations; haul roads will normally require additional equipment for maintenance purposes; and the number of water trucks or tankers must be determined.

Consider the portion of freeway relating to Figure 9.3 and Table 9.3, which is 87 stations in length. In addition to the 95400 CY to be over-hauled, as shown in column 2 of the table, there are $12560 + 10330 + 12623 = 35513$ CY to be excavated and placed within the limits of free haul at points *E, I* and *N*. Although the portion of freeway shown involves fairly heavy cut, it is only 1.6 miles long. Considering that typical freeway jobs average from 5 to 12 miles in length, it is clear that maximum efficiency must be attained on such work.

Using the quantities in Table 9.3, we have

Section	CY of cut	Stations of haul
1	7,000	84.8
2	23,100	62.7
3	41,200	20.5
4	17,500	17.2
5	6,600	11.9

Combining sections 3 and 4, whose haul lengths are about the same, we use the weighted average haul for the two sections, and have

Section	CY of cut	Stations of haul
1	7,000	84.8
2	23,100	62.7
3 + 4	58,700	19.5
5	6,600	11.9

The hourly rental and wage rates are assumed to be

Scraper, 21 CY struck, 30 CY heaped	$77
Bulldozer 270 HP	$60
Bulldozer 385 HP	$79
5 × 5 sheepsfoot roller	$ 7
Blade 125 HP	$39
Blade 150 HP	$40
Water tanker 5M gal.	$35
Grade foreman	$18.50
Grade checker	$16.10
Laborer	$13.48

Assuming the material to be a sandy loam, and that a 385 HP tractor will be used as a pusher, we conclude that the scrapers will carry $1.26 \times 21 =$

26.4 bank yards per load, which is equivalent to a load factor of 0.88 applied to the heaped capacity. From equation (9.3) we have

$$L_2 = \frac{26.4}{25} = 1.1 \text{ minutes}$$

and from equation (9.6)

$$U = \frac{26.4}{90} = 0.3 \text{ minutes}$$

The above times will be constant for all of the excavation, and are based on a 60-minute hour.

Since sections 3 + 4 contain the largest portion of the excavation, the scraper cycle time is estimated for this section first. Assuming essentially level haul roads, we have, from equation (9.4)

$$S = 445 + 0.114 \ (19.5 \times 100 \times 2) = 890 \text{ feet per minute}$$

and from equation (9.5) we find

$$H = \frac{19.5 \times 100 \times 2}{890} = 4.4 \text{ minutes}$$

The cycle time for section 3 + 4, based on a 60-minute hour is

$$1.1 + 4.4 + 0.3 = 5.8 \text{ minutes}$$

Based on a 50-minute hour, the production rate, per scraper, will be

$$\frac{50}{5.8} \times 26.4 = 227 \text{ BCY per hour.}$$

From equation (9.9) the cycle time for the pusher is found to be

$$P_2 = \frac{26.4}{18} = 1.5 \text{ minutes}$$

and the maximum number of scrapers that can be loaded with the one pusher is, from equation (9.7),

$$N = \frac{5.8}{1.5} = 3.9$$

If three scrapers are used, the production rate for sections 3 + 4, based on a 50-minute hour, will be

$$3 \times 227 = 681 \text{ CY/hr}$$

or

$$681 \times 8 = 5450 \text{ CY per 8-hour day.}$$

The time required to complete sections 3 + 4 will be

$$\frac{58700}{5450} = 11 \text{ days.}$$

By increasing the cycle time for the scrapers to 6.0 minutes, four scrapers may be used, but there will be some waiting, in the cut, for the pusher. Based on a 50-minute hour, the production rate for four scrapers will be

$$\frac{50}{6.0} \times 26.4 \times 4 = 880 \text{ CY/hr}$$

or

$$880 \times 8 = 7040 \text{ CY per 8-hour day}$$

and nine days will be required for completion.

Using the same procedures, the cycle times and the production rates are determined for sections 2, 1, and 5, and the data are entered in Table 9.4.

In addition to sections 1 through 5, there are 35513 CY of material within the free haul limit, as shown on the mass diagram (Figure 9.3). This quantity should be increased by the amount of the sidehill excavation taken from the road profile, and for this example we assume the total material, within the free haul limit, including such sidehill excavation, is 36500 CY.

As previously discussed, the effective production rate for rubber-tired scrapers working within the free haul limit is 11 cycles per hour per scraper. Thus, the effective (50-minute hour) cycle time is

$$\frac{50}{11} = 4.54 \text{ minutes}$$

Table 9.4. Possible Production Rates

Section	BCY	Number of Scrapers	Daily Production BCY[b]	Days to Complete
1	7000	3	3270	2.1
		4[a]	4350	1.6
		5	5440	1.3
2	23100	3	3910	5.9
		4[a]	5220	4.4
		5	6520	3.5
3 + 4	58700	3[a]	5450	10.8
		4	7040	8.3
5	6600	2[a]	4490	1.5
		3	6740	1.0
Free haul	36500	2[a]	4650	7.8

[a]Number of scrapers selected.
[b]Based on a 50-minute hour and an 8-hour day.
Note: Estimated production rates (BCY per day—see text):

Double 5×5 sheepsfoot roller	5600
Twin vibratory sheepsfoot roller	6400
5000 gallon water tanker	3200

and the equivalent cycle time based on a 60-minute hour would be

$$4.54 \times \frac{5}{6} = 3.78 \text{ minutes.}$$

As previously determined, the pusher cycle time, based on a 60-minute hour is 1.5 minutes, and, from equation (9.7) we find

$$N = \frac{3.8}{1.5} = 2.5.$$

Two scrapers will be used, and the effective production rate, based on a 50-minute hour, will be

$$2 \times 26.4 \times 11 = 581 \text{ CY/hr}$$

or

$$581 \times 8 = 4650 \text{ CY per 8-hour day,}$$

and eight days will be required to complete this work. This information is entered in Table 9.4, where we now have an array of the possible production rates, by sections, for various combinations of equipment. Such an array will simplify the final selection of the number of scrapers to use for each section, since it may not be possible or feasible to bring in an additional scraper for a short period of time. Before making this decision, the production rates for the compacting and watering equipment should be considered.

The production rates for a double 5×5 sheepsfoot roller and for a twin vibratory sheepsfoot roller are entered under Table 9.4.

9.1.7.1 WATER REQUIRED FOR COMPACTION.   The amount of water required for compaction is usually estimated from past job records. In the absence of such records an approximation may be made. Assume the following information is known:

Maximum density of the soil	113 lb/CF
Optimum moisture content	10%
Density at 90 percent relative compaction	101.7 lb/CF
Moisture in bank material	2%
Grading factor	0.80

The moisture percentages are determined on a weight basis. The theoretical amount of added water required for compaction is $10 - 2 = 8.0$ percent. Adding 1 percent for field compaction, and 5 percent for waste, evaporation, and for the water used to maintain haul roads, we estimate that 14 percent of additional moisture will be required. This amounts to

$$0.14 \times 101.7 = 14.2 \text{ pounds per compacted cubic foot}$$

or

$$\frac{14.2 \times 27}{8.33} = 46.0 \text{ gallons per compacted cubic yard}$$

which is

$$46.0 \times 0.80 = 37 \text{ gallons per bank yard.}$$

Figure 9B Semi-trailer type sprinkler tank, 10,000 gallons. (Southwest Welding and Manufacturing Co., Alhambra, California).

The cost to apply water is based on the following assumptions: The water supply is located near the middle of the job and the cost to develop and to furnish the water supply is included under a separate job item. Semi-trailer sprinkling tankers, having a capacity of 5000 gallons, will be used. The water will be loaded from a portable water tower having a 12000 gallon tank and an effective discharge rate of 2500 gallons per minute. Finally, we assume water tankers will discharge at a maximum effective rate of 1000 gallons per minute.

Assuming a round trip haul distance of 1.6 miles and an average tanker speed of 12 miles per hour, the cycle time for the 5000 gallon tankers is estimated as follows:

Loading time $\dfrac{5000}{2500} + 1$ $= 3.0$ minutes

Hauling time $\dfrac{1.6 \times 60}{12}$ $= 8.0$ minutes

Discharge time $\dfrac{5000}{1000}$ $= \underline{5.0}$ minutes

Minimum cycle time   16.0 minutes

and we estimate the effective cycle time will be 20 minutes per load.

Each tanker will therefore deliver 15000 gallons per hour, which is sufficient to compact

$$\frac{15000 \times 8}{37} = 3200 \text{ bank yards per 8-hour day.}$$

Figure 9C Hydraulically erected mobile water tower. (Southwest Welding and Manufacturing Co., Alhambra, California).

The production rates for the water tankers are noted at the bottom of Table 9.4.

Reviewing the information in Table 9.4, the following equipment is selected.

Section 1	4 scrapers
Section 2	4 scrapers
Sections 3 + 4	3 scrapers
Section 5	2 scrapers
Freehaul	2 scrapers

Two water tankers will be ample for all sections, and the material will be compacted with a large blade and a double 5×5 sheepsfoot roller.

9.1.7.2   MOTOR GRADERS REQUIRED ON THE FILL.   The number of motor graders required on the fill is usually based on the estimator's judgment and past experience. A rational approach, however, may be made as follows:

From Table 9.4, the maximum production rate is

$$\frac{5450}{8} = 681 \text{ BCY per 50-minute hour.}$$

Assuming a load factor of 0.85, this is equivalent to

$$\frac{681}{0.85} = 801 \text{ LCY per 50-minute hour.}$$

Assuming the material will be spread about 4 inches deep by 12 feet wide, we have

$$\frac{0.33 \times 12 \times L}{27} = 801 \text{ LCY}$$

and

$$L = 5500 \text{ feet or 55 stations}$$

where

$$L = \text{the length of the material spread per 50-minute hour.}$$

From equation (7.10), assuming an average speed of 4.5 miles per hour, and that three passes will be required to mix the water with the material, we have

$$P_S = \frac{4.5 \times 52.8}{3} = 79 \text{ stations per 60-minute hour}$$

$$= 79 \times \frac{50}{60} = 66 \text{ stations per 50-minute hour}$$

and the number of motor graders required would be

$$\frac{55}{66} = 0.8, \text{ and one scraper will be used.}$$

9.1.7.3 UNIT COSTS. The daily cost for the loading equipment is

385 HP Bulldozer 8 × $79 = $632,

the daily cost for the hauling units is

2-scraper string 2 × 8 × $77 = $1232
3-scraper string 3 × 8 × $77 = $1848
4-scraper string 4 × 8 × $77 = $2464

and the cost for the compacting equipment is

Blade 150 HP $40 per hour
Bulldozer 270 HP $60 per hour
Sheepsfoot roller $ 7 per hour
Two 5000 gal. tankers $70 per hour
                               ─────────────
                                $177 per hour

or

$$8 \times \$177 = \$1416 \text{ per 8-hour day.}$$

In all cases, the labor will consist of one grade foreman, two grade checkers, one in the cut and one on the fill, plus one laborer, who will be used to keep a load count and a cycle time record for the scrapers. The daily cost for this crew will be

$$8(18.50 + 2 \times 16.10 + 13.48) = \$513.44.$$

The unit costs for labor and for equipment are:

Section 1

$$\text{Labor} \quad \frac{513.44}{4350} \quad = \$0.118 \text{ per BCY}$$

$$\text{Rental} \quad \frac{632 + 2464 + 1416}{4350} = \$1.04 \text{ per BCY}$$

Section 2

$$\text{Labor} \quad \frac{513.44}{5220} \quad = \$0.098 \text{ per BCY}$$

$$\text{Rental} \quad \frac{632 + 2464 + 1416}{5220} = \$0.864 \text{ per BCY}$$

Sections 3 + 4

$$\text{Labor} \quad \frac{513.44}{5450} \qquad = \$0.094 \text{ per BCY}$$

$$\text{Rental} \quad \frac{632 + 1848 + 1416}{5450} = \$0.715 \text{ per BCY}$$

Section 5

$$\text{Labor} \quad \frac{513.44}{4490} \qquad = \$0.114 \text{ per BCY}$$

$$\text{Rental} \quad \frac{632 + 1232 + 1416}{4490} = \$0.731 \text{ per BCY}$$

Freehaul

$$\text{Labor} \quad \frac{513.44}{4650} \qquad = \$0.110 \text{ per BCY}$$

$$\text{Rental} \quad \frac{632 + 1232 + 1416}{4650} = \$0.705 \text{ per BCY}$$

9.1.7.4   MAINTAINING THE HAUL ROAD.   The cost to maintain the dirt haul roads is estimated as a separate lump sum item on the estimate sheet for roadway excavation. Typically, the equipment allowed for this will be based on past job experience.

Referring to Table 9.4, we note that the excess watering capacity available from the fill tankers working on section 1 is

$$(6400 - 4350) \times \frac{37}{8} = 9.5 \text{ M-Gallons per hour.}$$

The excess watering capacity available is determined for the remaining sections, and the following array is set up.

Section	Excess capacity (M-Gal/hr)	Length of haul road (Sta)
1	9.5	85
2	5.5	63
3 + 4	4.4	20
5	8.8	12
Freehaul	8.1	10

The haul road lengths above are taken directly from column 5 of Table 9.3, except that the weighted average is used for sections 3 + 4.

Assuming, as before, that each haul road water truck will make three trips per hour, it will spend at least $3 \times 5 = 15$ minutes per hour, watering at the road. The excess capacity noted above for section 1 amounts to two loads per hour, or an additional 10 minutes of watering time per hour. Assuming an average speed, while watering, of 10 miles per hour, the watering equipment will cover

$$\frac{25}{60} \times 10 \times 52.8 = 220 \text{ Station/hour},$$

which is more than two passes per hour over the haul road for section 1. Considering the remaining sections in a similar manner, we find that one haul road water truck plus the available excess capacity from the fill will result in more than two passes per hour on section 2, that the excess capacity from the fill will result in 1.9 passes per hour on sections 3 + 4, and that the excess capacity from the fill is ample for the remaining two sections.

If one 3 M-Gal water truck is used on the haul roads, the total water available for section 1 is

$$3 \times 3 + 9.5 = 18.5 \text{ M-Gal/hr}$$

or

$$\frac{18500}{7.48} = 2470 \text{ CF/hr}$$

Assuming the haul road is 36 feet in width, we have

$$85 \times 100 \times 36 \times t = 2470$$

$$t = 0.008 \text{ ft/hr}$$

where

t = depth of water applied per hour; the weight of the water applied per hour is

$$0.008 \times 62.4 = 0.5 \frac{\text{lb}}{\text{hr}} \text{ per SF}$$

Assuming the soil weighs 100 lb/CF, a 6-inch layer will weigh 50 pounds, and we will be applying 1 percent, by weight, of moisture per hour, which will be adequate. Checking the remaining sections in a similar manner, we decide to use one 3000-Gal. water truck on the haul roads for sections 1 and 2. Sections 3 + 4, section 5, and the freehaul section will be maintained with the excess capacity from the fill tankers. Referring to Table 9.4, the water truck will be required for seven days.

One motor grader, traveling 4.5 miles per hour with an effective blade width of 10 feet, will make one pass over

$$4.5 \times 5280 \times \frac{10}{9} \times \frac{5}{6} = 22000 \text{ SY per 50-minute hour,}$$

and we will need a minimum of one pass per hour to maintain the haul road. Considering the haul road areas, we decide to use two 125-HP blades on sections 1 and 2, and one blade on sections 3 + 4. The remaining two sections will be maintained by the blade working on the fill. The days of blade-time required, from Table 9.4, are

$$(2 + 5) 2 + 11 = 25 \text{ days.}$$

Assuming hourly rental rates of $30 for the water truck and $39 for the motor grader, the equipment cost to maintain the haul roads is:

$$7 \times 8 \times 30 + 25 \times 8 \times 39 = \$9500.$$

9.1.7.5   ON AND OFF CHARGES FOR EQUIPMENT.   The determination of the cost to move the equipment on and off the job is the final item on the estimate sheet for roadway excavation. In addition, an estimated allowance is made for the survey stakes required.

9.1.8   Compacting the Original Ground

Specifications for roadway excavation normally require a relative compaction of 95 percent to be obtained for a minimum depth of 2 to 3 feet below the finish roadway grade, for at least the width of the traveled way, whether in excavation or in embankment. Thus, in all cut areas, and in areas of low embankments, undercutting (additional roadway excavation) is required.

The undercutting is made to a plane 8 inches above the lower limit for the compacted zone, and the unit costs derived for roadway excavation within the freehaul limit will apply to this work. The compaction of the bottom 8 inches is designated as compacting the original ground, and the cost for this work is subject to considerable variation from job to job.

The unit cost for the equipment required to compact original ground is indexed to the total hourly rental rate for the following equipment:

1 bulldozer
1 double 5×5 sheepsfoot roller
1 motor grader
1 water truck

The production rate achieved by this equipment will vary with the sizes of the individual areas involved, the type of material, and the degree of difficulty encountered in mixing the water with the material when it is dry, or in aerating and drying it out when it is too wet.

Areas of 30 MSY in sandy loam have been compacted at an effective rate of 0.70 MSY per hour, areas of 20 MSY in pure sand have been compacted at an effective rate of 1.2 MSY per hour, and areas of 2 MSY in sandy loam have been compacted at an effective rate of 0.47 MSY per hour. In most cases, no additional labor will be required, since the work will be done concurrently with the other roadway excavation.

9.2   Imported Borrow

Imported borrow is normally bid as a separate job item. In every case, the cost of the borrow pit site, or the royalty paid for the borrow material, should be the first item on the estimate sheet for imported borrow.

The unit price paid for imported borrow includes the costs of furnishing, loading, hauling, placing, and compacting the material, as well as the cost to make the grade on the completed embankment. The unit of measure, for payment, is usually bank yards excavated, or tons. Some agencies will, however, pay for the item on a compacted embankment yard basis.

When off-highway hauling is possible, and when the one-way haul distance is about 1.25 miles or less, scrapers are normally used. When the import must be hauled over public roads, or when the one-way haul distance is about 1.50 miles or more, the material is hauled in trucks.

Significant savings are possible when borrow pits can be located close to their related embankments, and jobs are often won by obtaining optimum pit sites. Where land values are high, deeper pits are used. Typical pits range from 10 to 20 acres in area and from 25 to 40 feet in depth.

Prior to making a final decision on a pit site, test holes are drilled to see if the proposed material meets the specifications. When occasional thin layers of unsuitable material are found, a satisfactory product can sometimes be obtained by opening up the pit and loading the remaining material from top to bottom. The material is thus loaded in a downhill direction, which is advantageous, and, by cutting across the layers, a mixed product is obtained at no additional cost.

The cycle time for hauling units working out of borrow pits is

$$T = L + P + S + H + U \tag{9.10}$$

where

T = cycle time in minutes.
L = loading time in minutes.
P = time required to get in and out of the pit in minutes.
S = time spent on scales in minutes.
H = haul time in minutes.
U = unloading time in minutes.

Scales, when used, are located at the pit exit, and the weighmaster is normally furnished by the owner at no cost to the contractor. The time spent on the scales will be about 0.2 minutes per load.

Scales may be rented by the month, and the labor cost to set up and remove a set of scales will be about $1500 based on $K_{6+2} = 1000$, and the equipment cost will be about $750 based on $K_{Cr} = 75$. In addition, from 5 to 7 cubic yards of concrete will be required, and the truck time for hauling the scales on and off the job will be additional.

Careful consideration must be given to the time required to get in and out of the pit. This time will depend on the pit size, its depth, the type of the pit material, the pit-ramp grade, and the type of hauling units used.

Shallow pits pose no problem for scrapers, but the haul road in the pit must be maintained and its length included when estimating the hauling time for the scrapers. Shallow pits may require lower truck speeds when the pit material is sandy or otherwise unsuitable for truck travel. In addition, when trucks are used, the pit haul roads must be maintained even though the out-of-pit haul is over public roads.

The average haul speeds for equipment working in deep pits with ramps will be about 400 feet per minute for scrapers and about 10 to 12 miles per hour for trucks.

Except for the additional consideration of the possible loss of time in the pit, the procedures for estimating the costs for borrow pit excavation with scrapers are the same as those illustrated previously for roadway excavation. When trucks are used, the unloading time requires careful analysis, and will include the time spent hauling over the embankment in progress.

Although haul roads over original ground or completed embankments can be well maintained, the top layer of the fill in progress will be composed of loose or partially compacted material which will result in lower truck speeds. To cut the unloading time, sets of doubles are normally used to haul import when large quantities are involved. These consist of a truck-tractor pulling two bottom-dump trailers of equal size, the front trailer being a semi-trailer. Such units have a shorter turning radius than a semi of the same capacity, can hit the fill at high speeds, and can dump on the fly. When import is hauled in end-dump trucks, an additional 3 to 5 minutes per load is required for the dumping time.

The estimation of the time spent on the fill requires job experience, and the time will depend on the length and grade of the fill, the type of material being placed, and the possible approach speeds.

Wheel loaders are frequently used to load trucks with imported borrow. When high production rates are necessary, large loaders are required, and these are seldom available for rental.

9.2.1 Haul Distances for Imported Borrow

The determination of haul distances for imported borrow is usually simple, since all the material from a given pit will normally go into one or two large embankments. The centers of volume for the embankments are often determined by inspection, since the embankment quantities are tabulated at uniform intervals along the road profile. When necessary, centers of volume for embankments are determined by the methods illustrated in Section 9.1.2.

When a single borrow pit is located at the side of a road job requiring import along its entire length, haul distances must be determined by dividing the job into two sections, the point of division being the point where the haul road from the pit enters the job right-of-way. The haul lengths will then be the distances, measured along the haul roads, from the center of volume of the pit to the centers of volume of those embankments located along each of the two job sections. In such cases, it is incorrect to assume that the average haul length for all the import is the distance from the center of volume of the pit to the center of volume of all of the embankments on the entire job.

Figure 9D Kolman belt loader, 5 foot belt-50 feet long. (Kolman Division, Athey Products Corp., Sioux Falls, South Dakota.

9.2.2 Belt Loaders

Specially designed conveyor belts, fed by two to four bulldozers, are widely used when large amounts of earth must be loaded into trucks. The belt loader is a semi-portable, self-contained unit from 50 to 60 feet long, with a belt width of 4, 5, or 6 feet. The loader is equipped with a low profile dozer trap and a heavy duty plate feeder. When needed, a vibrating screen or grizzly may be added to scalp off oversize material, and this important capability is unique to the belt loading method.

Provision must be made, in all cases, to keep the truck loading area free of spillage or scalped material. A small rubber-tired tractor equipped with a back blade is used to remove spillage, since it is cheaper and more maneuverable than a motor grader.

The output capacity of the belt loader depends on the number of bulldozers used to feed it, the capacity of the belt, the size of the haul units, and the ability of the belt operator, who must stop the belt after each haul unit is loaded. Assuming the belt loader itself has ample capacity, the output of the system may be increased by as much as 33 percent for short periods of time. This is accomplished by decreasing the haul distance for the bulldozers, and is a second important and unique characteristic of the belt method of loading. A belt loader can get a string of waiting trucks loaded and back on their normal cycles in less time than is required by any other method of loading.

Belt loaders are moved back into the cut as the work progresses. This is done after the normal work shift; two bulldozers and two workmen will make the move in about 2 hours.

9.2.2.1 BELT LOADER EXAMPLE. The following conditions are assumed:

1. The pit is an average of 600 feet × 1000 feet in area by 25 feet deep, and contains 550000 bank yards. The material is sandy loam weighing 1.50 tons per bank yard.
2. The haul road is 3.2 miles long, one way, from pit exit to the beginning of the fill, and no public roads are used.
3. Average haul distance on the fill is 900 feet.
4. The material is hauled in sets of doubles having a capacity of 24 tons per set.
5. Scales are not required.

Based on past job records, a belt loader with a 5-foot-wide belt will load 24 tons of material into a set of doubles in 1 minute, which is equivalent to 1440 tons of material per 60-minute hour. Thus, referring to equation (9.10), we have

$$L = 1.0 \text{ minutes.}$$

One 270-HP bulldozer with a straight blade should move about 340 LCY per 60-minute hour on a 150 foot haul. Assuming a load factor of 0.80, this is equivalent to

$$340 \times 0.80 = 270 \text{ bank yards per hour}$$

or

$$270 \times 1.5 = 400 \text{ tons per hour.}$$

Four 270-HP bulldozers will be used. The estimated production rate for the four bulldozers is 1600 tons per 60-minute hour.

Assuming the haul road is well maintained and has no excessive grades or sharp turns, the average haul speed for the doubles will be about 27 miles per hour. Referring to equation (9.10), we have

$$H = \frac{2 \times 3.2 \times 60}{27} = 14.2 \text{ minutes, based on a 60-minute hour.}$$

Assuming the trucks average 10 miles per hour on the fill, and that they dump on the fly, we have

$$U = \frac{2 \times 900 \times 60}{5280 \times 10} = 2.1 \text{ minutes, based on a 60-minute hour.}$$

Figure 9E Kolman belt loader. Screening oversize material and loading bottom-dump semi-trailers. (Kolman Division, Athey Products Corp., Sioux Falls, South Dakota).

Assuming the in-pit haul to average about 600 feet, one way, and an average truck speed of 12 miles per hour, we have

$$P = \frac{2 \times 600 \times 60}{5280 \times 12} = 1.1 \text{ minutes, based on a 60-minute hour.}$$

From equation (9.10) the total cycle time, based on a 60-minute hour, is

$$T = 1.0 + 14.2 + 2.1 + 1.1 = 18.4 \text{ minutes,}$$

and the production rate, per set of doubles will be

$$\frac{60}{18.4} \times 24 = 78.3 \text{ tons per 60-minute hour.}$$

The number of trucks required is

$$\frac{1440}{78.3} = 18.4 \text{ trucks,}$$

and 19 trucks will be used. Both the production rates used to determine the number of trucks required are based on 60-minute hours.

The effective production rate, based on a 50-minute hour, will be

$$1440 \times \frac{50}{60} = 1200 \text{ tons per hour,}$$

which, in this case, is equivalent to

$$\frac{1200}{1.50} = 800 \text{ bank yards per hour.}$$

Based on the pit size and the production rate, we conclude that the belt will be moved back into the pit face on 5-day intervals, which is equivalent to one move for each 48000 tons.

The equipment required to water and compact the fill, and the equipment required to maintain the haul road, are determined by the procedures outlined previously in the example for roadway excavation.

The labor required for the entire operation will be one grade foreman, one grade checker in the pit, one grade checker on the fill, and one laborer at the belt loader, who will keep a log of the truck cycle times.

The unit costs for imported borrow are determined as shown in the preceding example for roadway excavation. Since only one haul road is involved, the equipment cost to maintain the haul road may be based on a unit cost per ton of material hauled.

Thus, if the rental cost for the haul road equipment were $144 per hour, the unit cost for haul road maintenance would be

$$\frac{144}{1200} = \$0.12 \text{ per ton}$$

based on the effective production rate of 1200 tons per hour. Since the total length of the haul road, including the section in the pit, is 3.3 miles, we have an hourly haul of

$$1200 \times 3.3 = 3960 \text{ ton-miles}$$

and, for comparison when estimating future work, the haul road maintenance cost might also be expressed as

$$\frac{144}{3960} = \$0.0364 \text{ per ton-mile.}$$

9.3   Making the Grade

The cost to make the final grade will depend on the degree of care exercised in making the rough grade for the area being considered and on the allowable tolerances for the final grade. The rough grade for the roadbed in cut or embankment should be made to about ±0.3 foot at the time the work is done, and care should be taken to balance the plus and minus areas, with any error on the plus side.

The base unit cost for the equipment required to make grade is, for estimating purposes, indexed to the hourly rental costs for the following equipment, and these rental costs are assumed as shown.

1 blade	$39
1 water truck	$30
1 self-propelled pneumatic roller	$36
½ of 9 CY elevating scraper $46 × ½	$23
	$128

The sum of the hourly rental costs, above, is designated as K_{MG-R}.

The base unit cost for the labor required to make grade is, for estimating purposes, indexed to the hourly wage rates for the following crew, and these wage rates are assumed as shown.

1 grade foreman	$18.50
1 grade checker	$16.10
1 guinea hopper	$13.70
1 laborer	$13.48
	$61.78

The sum of the hourly wage rates, above, is designated as K_{MG-L}. On jobs where the quantity of grade to be made amounts to about 16 MSY or more, the above equipment and crew will make grade on the roadbed at an effective rate of about 1 MSY per hour, assuming tolerances of ±0.1 foot on the shoulders, and ±0.05 foot on the subgrade.

Thus, based on $K_{MG-R} = 128$, the base unit equipment cost for making grade on the roadbed is $128 per MSY, and, based on $K_{MG-L} = 61.78$, the base unit labor cost is $61.78 per MSY. These unit costs assume all areas are large enough to permit free and unrestricted use of the equipment.

Lower production rates are realized when smaller areas are involved. Thus grade will be made on areas of 3 MSY at a rate of about 0.45 MSY per hour, and on areas of 10 MSY the rate will be about 0.65 MSY per hour. When areas as small as 0.3 MSY are involved, the production rate will be about 0.35 MSY per hour.

When areas are confined and provide a minimum of space for the equipment, the unit costs for making grade increase, and the estimated amount of the increase will depend on the estimator's judgment.

The grade tolerance for medians is normally ±0.2 foot for the general area, and ±0.1 foot for drainage inverts. The crew and equipment listed above will make grade on median areas of 16 MSY or more at an effective rate of about 1.5 MSY per hour.

10

Base Courses and Pavements

Aggregate subbase, aggregate base, and cement treated base are commonly used as base courses in the structural sections of roadways. The thicknesses and types of base courses required are shown on typical roadway cross sections such as the one illustrated in Figure 9.1.

The unit price paid for the base course includes the cost to furnish, spread, and compact the material, and the cost to make the grade on top of it. The payment for base courses is based on the plan dimensions for the structural section. Thus, when material is paid for on a cubic yard basis, the unit of measure is compacted cubic yards.

Portland cement concrete pavement may be placed by general contractors, but is often installed on a subcontract basis. Asphalt paving is normally installed by subcontractors who have permanent mixing plants and aggregate sources in the area.

10.1 Aggregate Subbase (AS)

Aggregate subbase is a sandy material having a specified gradation, and, when required, is the bottom course in the structural section. Aggregate subbase is obtained from borrow pits, commercial rock plants, or, in some instances, from the roadway excavation. When obtained from borrow pits, the costs to furnish the material to the grade are estimated by the methods outlined in Chapter 9, except that the dumping area will be relatively restricted, and lower production rates will be realized.

Typical sections of aggregate subbase will range from 0.5 feet to 1.5 feet in thickness. The material is dumped on the grade, spread with a motor grader, and compacted to a relative compaction of 95 percent. When the section is more than 0.5 feet in thickness, the specifications may require the material to be placed in two or more layers of approximately equal thickness, and the compacted thickness of an individual layer, in such cases, cannot exceed 0.5 feet. Typically, a small wheel tractor pulling a set of discs will be required to mix the water with the subbase during compaction.

When the material can be hauled with scrapers, a top production rate of 300 compacted cubic yards per hour may be maintained on large jobs where the structural section is about 50 feet wide or more. The costs for hauling with scrapers were discussed in Chapter 9.

When the material is hauled by trucks, the maximum production rate will be about 300 tons per hour, and this rate may not be practicable where the hauls are long, because of the number of trucks required. When aggregate subbase, in relatively small amounts, is furnished by commercial plants, the loading may be done with plant equipment, and a total of about 10 minutes should be allowed for loading and unloading the trucks.

When large amounts of material are required from commercial plants, the contractor should do his own loading or get a commitment from the plant concerning the loading rate, if the plant loaders are used. In such cases the truck cycle times are estimated as shown in section 9.2.2. In general, whenever trucks are used, a load counter should keep a log of the truck cycle times, and when large quantities are involved, a truck foreman should be used to keep the trucking operation running efficiently.

A grade checker is used on the grade to spot the trucks for dumping. Although aggregate subbase may be moved about during the spreading and compacting operation, savings are achieved when the dumping is controlled to insure that the tonnage dumped per linear foot of roadway is approximately equal to the theoretical amount required. The procedures used to determine the dumping rate are the same as those outlined in the following section for aggregate base, except that somewhat less precision is necessary.

10.1.1 Spreading and Compacting Aggregate Subbase

The unit costs to spread and compact aggregate subbase depend on the delivery rate for the material, and it is imperative that the delivery rate assumed on the estimate be achieved or bettered during the course of the work. When suppliers have quoted the material F.O.B. the job, the contractor must satisfy himself that the required delivery rate will be realized with the loaders and trucks which the supplier proposes to use.

A typical string of equipment required to spread and compact aggregate subbase is:

1 motor grader
1 185-HP bulldozer pulling
1 vibratory steel roller (8000 lbs.)
1 small rubber-tired tractor pulling a set of discs
1 or more water trucks

and the labor required will be

1 grade foreman
1 grade checker

The number of water trucks used in the estimate is based on past job experience, or is estimated as shown in Section 9.1.7.1. Aggregate subbase is frequently quite dry, and it is essential to give careful consideration to the amount of water required.

The unit costs for spreading and compacting aggregate subbase are determined by assigning the appropriate hourly rental and wage rates to the equipment and labor shown above, and the resulting total hourly costs are divided by the estimated delivery rate for the typical job situation. The resulting base unit costs are increased (or the delivery rate used is decreased) for those sections which are more confined, and the amount of this adjustment will depend on the estimator's judgment. The unit costs for dumping, spreading, and compacting aggregate subbase will be the same regardless of the number of layers required.

10.1.2   Making the Grade on Aggregate Subbase

It is necessary to make the grade only on the top layer of aggregate subbase, since the approximate grades for the layers below are achieved, at no additional cost, by controlling the rate of dumping.

The base unit costs for making grade are indexed to the same equipment and crew used to make grade on earthwork (section 9.3) and the production rate will be about 0.80 MSY per hour, when the grade tolerance is ± 0.05 foot. Thus, based on $K_{MG-R} = 128$, and $K_{MG-L} = 61.78$, the base unit costs for equipment and labor will be:

Equipment: $\dfrac{128}{0.80}$ = $160 per MSY

Labor: $\dfrac{61.78}{0.80}$ = $77.20 per MSY

The above base unit costs assume quantities of 10 MSY or more, and free access for the equipment.

10.2   Aggregate Base (AB)

Aggregate base is normally placed immediately above the aggregate sub-base, and the thickness of the aggregate base section will vary from about 0.33 to 1 foot. The material is usually furnished by a commercial rock plant, although, on large jobs, it may be manufactured at the site.

Typical gradations for aggregate base are as follows:

Sieve Sizes	Percentage passing	
	1½″ maximum	¾″ maximum
2″	100	—
1½″	87–100	—
1″	—	100
¾″	45–90	87–100
No. 4	20–50	30–60
No. 30	6–29	5–35
No. 200	0–12	0–12

Specifications normally require aggregate base to contain sufficient moisture for compaction at the time it is delivered to the grade. This requirement is imposed in an attempt to prevent undue segregation during the spreading and compacting operation.

10.2.1   Spreading and Compacting Aggregate Base

When the required thickness is greater than 0.5 foot, the material is spread in two or more layers of approximately equal thickness, and the compacted thickness of each layer cannot exceed 0.5 foot. Each layer must be spread in one operation, at a uniform rate per linear foot of roadway, and must be a full traffic lane in width.

Aggregate base is usually delivered in end-dump trucks, and dumped through a spreader box attached to the blade of a bulldozer. Some specifications allow the material to be dumped in windrows by bottom dump trucks. In such cases, the material is spread with a 150-HP motor grader that has end wings attached to the blade at the required width.

In all cases, the tons of material spread or dumped per linear foot of lane must be carefully controlled so that the compacted thickness of each layer will equal its required thickness. Typical specifications will not allow any appreciable amount of material to be moved about or shifted during the grade-making operation.

The compacted density of the moist aggregate base will range from 1.80 to 2.00 tons per compacted cubic yard, depending on the source of the material. When this density is not known, it is determined by laboratory tests.

The spreading or dumping rate for the material is determined as follows:

$$R = \frac{DWt}{27}$$

where

R = the spreading or dumping rate, in tons per linear foot of lane.
D = the density of the compacted moist material, in tons per cubic yard.
W = the width of the lane, or spread, in feet.
t = the thickness of the layer to be spread, in feet.

Thus, assuming D = 1.90 tons per cubic yard, W = 12 feet, and t = 0.5 foot,

$$R = \frac{1.90 \times 12 \times 0.5}{27} = 0.422 \text{ tons per foot}$$

and a truck containing 23.9 tons of material would be spread or dumped in a distance of

$$\frac{23.9}{0.422} = 56.6 \text{ feet.}$$

To control the spreading or dumping, the grade checker sets a temporary stake at this measured distance, and the distance required for the second load is measured from this stake, and not from the end of the previous spread or windrow.

The base unit equipment cost for dumping, spreading, and compacting aggregate base when a spreader box is used depends on the delivery rate for the material, and is indexed to the sum of the hourly rental rates for the following equipment:

1 150-HP bulldozer with spreader box
1 self-propelled pneumatic roller
1 motor grader
1 small (2000 gallon) water truck.

The base unit cost for labor is indexed to the sum of the hourly wage rates for the following workmen:

1 grade foreman
1 grade checker

The above equipment and labor will dump, spread, and compact aggregate base at a maximum rate of 300 tons per hour. To achieve this rate, free access must be available. On narrow ramps the rate will drop to about 150 to 200 tons per hour. In very tight areas, the rate will be still lower. 600 tons of aggregate base were dumped, spread, and compacted in a strip 2.5 feet wide by 0.5 foot deep, at a rate of 45 tons per hour. The unit cost for dumping, spreading, and compacting aggregate base is the same regardless of the number of layers required.

When the material is delivered in bottom dump trucks, an additional grade checker is necessary to spot and dump the trucks, and chains are used to control the gate openings. The equipment and labor needed to spread and compact the material are the same as that listed above, except that the bulldozer and spreader box are replaced with a 150-HP motor grader. Bottom-dumping aggregate base in windrows is advantageous when job conditions preclude fast delivery rates. The dumping is performed as a separate operation, and the increased cost due to the low delivery rate is small, since only one workman is involved. When sufficient material has been delivered to the grade, the spreading and compacting operation is done at the 300 tons-per-hour rate.

The costs to deliver the material to the grade are based on the same considerations discussed previously in connection with aggregate subbase. About 5 minutes are required to spot and dump each truck, whether through a spreader box, or in windrows. When the material is loaded by a commercial plant, at least 5 minutes must be allowed for loading each truck and, in estimating haul times, consideration must be given to the time lost while the trucks move through the plant area.

As an example, assume the aggregate base will be plant-loaded, and that the one-way haul distances are 4 miles through a residential area and 20 miles on good county roads. Assume the trucks will carry an average of 23.5 tons per load, will travel at 26 miles per hour in the residential area and 41 miles per hour on the county roads, and that 10 minutes will be required to load and dump. Assuming that the stockpile is close to the public road, the cycle time for the trucks will be:

$$\frac{8}{26} + \frac{40}{41} + \frac{10}{60} = 1.45 \text{ hours per load}$$

and each truck will deliver

$$\frac{23.5}{1.45} \times \frac{55}{60} = 14.9 \text{ tons per 55-minute hour}$$

Assuming a desired production rate of 300 tons per hour, the number of trucks required will be

$$\frac{300}{14.9} = 20.1$$

20 trucks will be used, and the hourly production will be $20 \times 14.9 = 298$ tons per hour. Assuming an hourly rental rate, per truck, of \$30 per hour, the unit equipment cost to haul the material will be

$$\frac{20 \times 30}{298} = \$2.01 \text{ per ton}$$

The foregoing calculation assumes a large job where two or more weeks of hauling are involved. In such cases it pays to bring in trucks, if necessary, from outlying areas. If the job were small and only a day or two of hauling were required, the number of trucks used would be based on the estimator's judgment of the number that would be available, locally, for the short duration of the work.

In such a case, it might be concluded that 10 trucks would be available; the effective production rate would then be

$$10 \times 14.9 = 149 \text{ tons per hour,}$$

and the costs to spread and compact the material would be increased.

On the large job, the haul labor would consist of a load counter and a truck foreman. On the short job, only a load counter would be used, and the job superintendent would oversee the trucking operation.

10.2.2   Making the Grade on Aggregate Base

It is necessary to make the grade only on the top layer of aggregate base. The approximate grades for the layers below are achieved at no additional cost by controlling the rate of spreading and dumping.

The base unit costs for making grade are indexed to the same equipment and crew used to make grade on earthwork and on aggregate subbase, and the production rate will be about 0.87 MSY per hour when the grade tolerance is ±0.05 foot, when the quantity involved is 3 MSY or more, and when there is free access for the equipment.

10.2.3 Prime Coats

The specifications may require applying a prime coat of liquid asphalt to the finished surface of the aggregate base course. The asphalt is paid for under a separate job item on a price per ton basis.

SC-250 liquid asphalt is commonly used for prime coats. The material is sprayed from insulated and metered distributor trucks at temperatures of 140 to 225 degrees Fahrenheit, and the material is applied at uniform rates of about 0.25 gallons per square yard. Typical distributor trucks carry a maximum of about 11 tons of asphalt, and must carry a minimum of about 4 tons to prevent undue cooling of the material in transit.

Liquid asphalt is quoted, by the suppliers, at a price per ton which includes furnishing the material and applying it at the specified rate. Although the contractor incurs no additional costs in connection with the item, the quotations must be carefully analyzed, since the actual cost per ton may be considerably higher than the base price quoted. This increased cost results from several factors which will vary with different suppliers, but which may include such items as minimum load charges, a flat handling charge when unused material is returned, a reduced price credited for returned material, and freight charges which may be on a per load or on a per ton hauled basis.

Average Weights and Volumes of Liquid Asphalt

Grade of Liquid Asphalt	Gallons per Ton at 60° F
70	253
250	249
800	245
3000	241

The volume-temperature relationship for SC, MC and RC liquid asphalts, grades 250 through 3000 is

$$V_T = V_{60} \times F$$

where

V_T = volume at T degrees Fahrenheit.
V_{60} = volume at 60 degrees Fahrenheit.
F = a multiplier which equals 1.0000 for $T = 60$, increases 0.00035 per degree over 60, and decreases by the same amount per degree under 60.

10.2.3.1 PENETRATION TREATMENT. 17 tons of SC-250 penetration treatment was sprayed on a roadside ditch at a rate of 0.25 gallons per square yard. The liquid asphalt was applied with a hand spray attached to an insulated distributor truck, at an average rate of 0.57 tons per hour.

10.3   Cement Treated Base (CTB)

Cement treated base, when used, is the top base course in the structural section for the roadway. The material may be plant-mixed or road-mixed, and the method of mixing will be specified. CTB courses vary from about 0.50 foot to 1 foot in thickness, and are spread in layers having a maximum compacted thickness of 0.50 foot. The material is compacted to a relative compaction of 95 percent, and the compacted density will range from 1.80 to 2.00 tons per cubic yard.

Large general contractors, specializing in roadwork, frequently place cement treated base as well as portland cement concrete pavement with their own crews and equipment. Much of this work is, however, done on a subcontract basis. Electronically controlled paving machines are used to spread and compact the material, which may be spread up to four lanes in width. Grade tolerance problems make it desirable to have the CTB course and the paving placed by the same subcontractor.

10.3.1   Plant-Mixed CTB

Plant mixing, when required, is usually done by separate subcontractors who set up portable mixing plants at the job site, and who will mix from 150 to 450 tons of CTB per hour. The cement, aggregates, and water are supplied to the plant by the general contractor, who may also deliver the mixed material to the grade.

The cement companies quote the cement, loaded into silos, at the mixing plant. The cement is paid for by the general contractor, and an allowance of about 5 percent is made for waste.

10.3.2   Road-Mixed CTB

The aggregates for road-mixed CTB are normally delivered to the grade by the general contractor in bottom dump trucks. The material is dumped at a carefully controlled rate per linear foot of windrow, and the procedures necessary to determine the dumping rate are the same as those described in connection with aggregate base. About 5 minutes are required to dump each truck, and a grade checker is needed to spot the trucks and control the dumping operation.

The windrows are then shaped to a uniform cross section with a sizing device which may be attached to a motor grader, and such shaping is normally done by the CTB subcontractor. Not more than 4 hours before mixing, the cement is metered onto the windrows from trucks furnished by the cement supplier; this operation does not entail any additional labor cost to the contractor, but 5 percent is allowed for waste cement.

The windrowed materials are then picked up and mixed by a road-mixing machine, and the water is supplied to the machine by the general contractor, who furnishes one or more water trucks for this purpose; from 6 to 7 percent, by weight, of water will be required.

10.3.3   Job Pits for CTB Aggregates

The aggregates for cement treated base are often produced from pits located near the job site. Oversize material must normally be removed, and belt loaders are used to load and separate the material.

A 385 HP bulldozer, together with a 48-inch belt loader equipped with a 4 by 9 foot vibrating single deck screen, loaded trucks at a rate of 475 tons per 60 minute hour. Approximately 6.5 percent of the pit material was scalped off, and the total production for the bulldozer and belt was 508 tons per hour. The 24-ton trucks averaged 23.7 tons per load, and a small wheel loader was used to clear the scalped material from the loading area.

10.3.4   Blending CTB Aggregates

The pits used for the job-production of CTB aggregates are frequently deficient in rock, which is the material retained on the No. 4 sieve. In such cases, additional rock must be imported, or some of the pit sand (material passing the No. 4 sieve) must be wasted. Separating the sand to be wasted is impractical for a job plant, so additional rock is imported.

A typical grading specification for CTB aggregates is shown on line 1 of Table 10.1. Line 2 shows the gradation of an assumed pit sample; 7.0 percent of the pit material is oversize, and must be wasted.

Line 3 shows the gradation of that fraction of the pit sample which passed the 1-inch sieve, and line 4 shows the gradation of the rock to be imported. The imported rock will be blended with the scalped pit material so that 70.0 percent of the combined product will pass the No. 4 sieve. This percentage will furnish a leeway, or safety factor, of 5 percent with respect to the specifications.

Table 10.1. Material Grading

Material	% Used	Sieves—% Passing					
		2"	1"	¾"	No. 4	No. 30	No. 200
(1) CTB specification		—	100	87–100	35–75	7–45	0–19
(2) Typical pit sample		100	93.0	—	—	—	—
(3) Scalped pit sample		—	100	98.1	80.8	36.0	9.2
(4) Imported rock		—	100	100	0	0	0
(5) Scalped pit	86.6	—	86.6	85.0	70.0	31.2	8.0
(6) Imported rock	13.4	—	13.4	13.4	—	—	—
(7) Combined product	100	—	100	98.4	70.0	31.2	8.0

Since 80.8 percent of the scalped pit product is sand (line 3, column 5), the percentage of rock is $100.0 - 80.8 = 19.2$ percent. Since 80.8 percent of the scalped pit product must equal 70.0 percent of the combined product, the following proportion is set up:

$$\frac{Rc}{19.2} = \frac{70.0}{80.8}$$

where

Rc = the percentage of rock in the combined product,

and

$$Rc = 19.2 \times \frac{70.0}{80.8} = 16.6 \text{ percent.}$$

These data are tabulated as follows:

Pit material	Percent of pit product	Percent of combined product
Sand	80.8	70.0
Rock	19.2	16.6
	100.0	86.6

The percentage of imported rock required is $100.0 - 86.6 = 13.4$ percent of the combined product.

The percentages of material in the combined product are shown on lines 5, 6, and 7 of Table 10.1. The percentages of scalped pit material shown on line 5 equal 86.6 percent of the percentages shown on line 3. Similarly, the percentages of imported rock shown on line 6 equal 13.4 percent of the percentages shown on line 4.

The combined product, shown on line 7, is the sum of lines 5 and 6, and meets the specified gradation.

When CTB is plant mixed, the imported rock is delivered to a separate stockpile at the plant and the blending entails no additional cost. When the CTB is road mixed, the pit material is dumped in a controlled windrow from bottom dump trucks, and the imported rock is dumped in a second controlled windrow, directly on top of the first one. The dry aggregates are then mixed with one or more passes of the road-mixing machine, the blended windrow is sized, and it is then ready for the cement.

Assume that a 4.5-cubic-foot windrow containing 0.30 tons of aggregates per linear foot is required, and that the materials shown on lines 5 and 6 of Table 10.1 will be used. The amount of pit material windrowed is

$$0.30 \times 0.866 = 0.260 \text{ tons per linear foot}$$

and the amount of imported rock required is

$$0.30 \times 0.134 = 0.040 \text{ tons per linear foot}$$

Assuming the material is delivered in 24 ton loads, the pit material will be dumped in a measured distance of

$$\frac{24}{0.26} = 92.3 \text{ feet}$$

and the imported rock trucks will be dumped in a measured distance of

$$\frac{24}{0.04} = 600 \text{ feet}$$

In this case, a grade checker will dump the pit trucks in about 5 minutes, but will require about 15 minutes to dump the imported rock trucks.

Figure 10A Slipform paving machine, (Owl Slipform Concrete Co.).

10.3.5   Curing CTB

The completed cement treated base is covered with a bituminous curing seal. MC-250 liquid asphalt is used, and is applied at a rate of from 0.15 to 0.25 gallons per square yard of surface. The procedure is identical to that described for placing the prime coat on aggregate base, and entails no additional labor costs for the general contractor.

10.4   Portland Cement Concrete Pavement

Portland cement concrete pavement is often placed by subcontractors. Sometimes the work is done by paving firms that do nothing but this type of work, and in other cases the pavement is placed, on a subcontract basis, by large general contractors who specialize in the construction of roads and highways.

The job batch plants and paving machines required are expensive, highly specialized, and not available for rental. Due to the high cost, this equipment is owned only by contractors who do large amounts of portland cement concrete paving. Highway contractors do this work on a subcontract basis in an effort to keep their expensive equipment working.

The general contractor furnishes the site for the plant itself, and from 5 to 7 acres are required. He also furnishes the water supply; about 350 gallons per minute are needed, assuming a production rate of 500 CY per hour. The general contractor usually furnishes the aggregates delivered to the job plant, and will, in all cases, furnish and maintain the necessary job haul roads. The cement is normally set up as a separate job item, and is sometimes furnished by the general contractor.

Cement companies quote the cement loaded into job silos, and normally furnish the necessary silos at no additional cost. The general contractor incurs no labor nor equipment costs in connection with the cement item, but an allowance of 5 percent is made for waste cement. The yield for the mix design used must be carefully checked, since large quantities of cement are required, and the yield should be re-checked as the work progresses.

The aggregates are delivered in bottom dump trucks, are dumped into receiving hoppers, and loaded onto stockpiles by conveyors. The equipment required for unloading is furnished and installed by the paving subcontractor. About 1.07 tons of rock and 0.66 tons of sand are required per cubic yard of concrete, including the allowance for waste.

Assuming a slipform paver is used, and that the batch plant is equipped with two 8-cubic-yard mixers, the maximum production rate will be about 500 cubic yards per hour. The mixed concrete is delivered to the grade in end dump trucks, which are rented by the paving contractor.

Assuming a production rate of 500 cubic yards per hour, about 36 loads of aggregates and 62 loads of mixed concrete are delivered per hour. At the same time, on an average job, aggregates for base courses are arriving, and the resulting truck traffic on the job site prevents the production rate from being increased efficiently, even though larger mixers might be used.

The aggregates are normally obtained from commercial rock plants and, in view of the quantities involved, the general contractor should load the trucks. In addition, a truck foreman and a load counter should be used to keep the operation running efficiently. It is essential that the job haul roads be well maintained, since large quantities of materials will be hauled, and the aggregate trucks travel at speeds up to 40 miles per hour in many instances. There are no labor costs in connection with haul road maintenance, except for the costs of flagmen, if required.

10.5 Asphalt Pavements

Asphalt concrete pavement and plant-mixed surfacing are typically placed by subcontractors, and their effective production rate ranges from 1800 to 2000 tons per day, when working in unconfined areas. The average in-place density of both types of asphalt paving is about 2 tons per cubic yard. The general contractor will incur no expenses in connection with the asphalt paving item except for specific incidental items of work which may be excluded in the quotation by the subcontractor.

Plant-mixed surfacing is sometimes used to form spillways on embankment slopes. Based on $K_L = 13.48$ and $K_{BH} = 25$, areas averaging 4 SY each were graded at unit costs of $6.00 per square yard for labor, and $1.50 per square yard for equipment. A 1½-inch-thick layer of plant-mixed surfacing was placed and compacted, by hand, on these areas, at a unit labor cost of $5.20 per square yard.

Based on $K_L = 13.48$ and $K_{MR} = 74$, plant mixed surfacing was placed over pipe trenches at unit costs of $12.50 per ton for labor and $5.10 per ton for equipment, where $K_{MR} = $ the sum of the hourly rental rates for one motor grader plus one self-propelled steel roller. About 18 tons of material were placed at each trench.

Based on $K_L = 13.48$ and $K_{BH} = 25$, manhole covers were raised and adjusted to grade, after paving, at unit costs of $124 each for labor and $3.20 each for equipment, and these prices included placing the necessary concrete manhole rings, and the cost to hand place surfacing at the raised manhole cover on completion.

11

Clearing and Grubbing—Finishing Roadway

Clearing and grubbing is the first item done, and finishing the roadway is the last. In both cases, road building equipment is used to do the work.

11.1   Clearing and Grubbing

Clearing and grubbing is usually bid at a lump sum price. Since the item is performed early and the quantity is fixed, clearing and grubbing is normally bid at a price considerably greater than the bare cost for doing the work. In many instances, the price bid for clearing and grubbing will be about 10 percent of the total job price.

Based on $K_L = 13.48$ and a rental rate of $39 per hour for a motor grader, areas ranging from 2000 to 5000 MSY, covered with sagebrush and light mesquite, have been cleared at average unit costs of about $2.50 per MSY for labor, and about $7.60 per MSY for equipment. The unit cost for labor represents the cost to burn the brush.

Trees greater than about 3 inches in diameter are cleared with bulldozers having a minimum of 250 HP, and when the trees are greater than about 12 inches in diameter, a 385 HP bulldozer with a machine-mounted ripper should be used.

The number of trees to be removed is based on an actual count, for small areas, and on the counted number of trees in a typical acre, when the areas are large. The time required to uproot trees is about as follows:

Diameter (inches)	Time (minutes)
4–8	3
12	6
24	15
30	20

The roots of the larger trees are cut with the ripper, and the trees are then pushed over. The times shown above include the time required for ripping, uprooting, and filling the resulting holes.

When large numbers of trees must be removed, subcontractors are normally used. Such subcontractors have the necessary equipment, skilled crews, and supervision, and often have a market for the trees they remove.

11.2   Finishing Roadway

Upon completion of all construction operations, the roadway must be cleaned up and finished as specified. Typical work required is:

1. Trim and finish all embankment and cut slopes.
2. Trim, shape, and recompact the shoulders, which have been disturbed during paving and slope-trimming.
3. Clean out drainage ditches and pipes as required.
4. Remove all weeds.
5. Sweep paved areas.
6. Clean up and remove all debris, and leave the site in a neat and orderly condition.

The finishing roadway item is usually bid at a lump sum price, and since the item is the last one completed, the price bid should be no greater than the estimated bare cost for doing the work. A review of the bid prices for the finishing roadway item indicates that it is frequently bid at about half of this amount.

11.2.1   Trimming Slopes

Slopes up to 12 feet high are trimmed with a motor grader; one grade checker and one laborer will be required. About 100 stations of slope on one side of the roadway can be trimmed per 8-hour day.

Slopes up to 30 feet in height are trimmed by dragging anchor or dredge chain with two pieces of equipment, and the chain used will weigh from 45 to 60 pounds per linear foot. A motor grader or a wheel loader is used on the grade, and a bulldozer is used on the natural ground. One grade checker and about four laborers will be required, and about 10 stations of slope on one side of the roadway can be trimmed per hour.

The cost to trim embankment slopes at bridge abutments depends on the accuracy of the slope stakes used and on the amount of care exercised in placing the fill. Embankment construction at bridge abutments should be carefully supervised, since in most cases 1½:1 slopes under the bridge are intersecting with 2:1 slopes at the sides. The slope stakes at the intersections must be set at the proper radius for each stake, and the grade checker must check the grade, during construction, on the radial to the stake being used.

The cost to trim the abutment slopes for a highway overcrossing having a width of about 36 feet may be estimated by allowing from 8 to 16 hours for a 270-HP bulldozer, one laborer, and one grade checker, per bridge. Skewed bridges with ramps require more time than square bridges with no ramps.

11.2.2   Trimming Shoulders

One motor grader, one self-propelled pneumatic roller, one-half water truck and one grade checker are required. About 30 stations of shoulder, up to 8 feet in width, will be backed up and trimmed per hour.

11.2.3   Cleaning Ditches and Culverts

The estimated cost for this work depends on the estimator's judgment in view of the particular job conditions. On most well run jobs where the downstream drainage has been completed, very little work will be required. An average allowance of about 2 hours of laborer time per culvert and per drainage inlet will cover the cost of whatever work may be required.

11.2.4   Remove Weeds

This work is done with a motor grader, and no additional labor is required where the weeds are small. The production rate will be about 10 MSY per hour.

11.2.5   Sweep Paved Areas

A power road broom is used and no additional labor is required. The production rate will be about 12 MSY per hour.

11.2.6   Clean-up

The incidental clean-up costs represent a substantial portion of the total cost for the finishing roadway item, and in general, these costs cannot be predicted with much accuracy at the time the estimate is being prepared. Based on $K_L = 13.48$ and $K_M = 39$, where K_M = the hourly rental rate for a motor grader, the incidental costs have averaged about $11 per station for labor and $6 per station for equipment on freeway jobs, in rural areas, where 1000 or more stations were involved.

11.2.7   Overall Unit Costs

The estimated cost for finishing the roadway is often based on cost records showing the total labor and equipment costs for similar jobs. Such total costs are converted to unit costs per station of roadway. The stations used include the stations of all ramps and frontage roads; when there is a divided highway, the stations of both sides are included in the total. Thus, if a given project consists of 150 stations of divided highway, 70 stations of ramps, and 170 stations of frontage roads, the unit costs would be based on 150 + 150 + 70 + 170 = 540 stations of roadway.

Based on $K_L = 13.48$ and $K_M = 39$, where K_M = the hourly rental rate for a motor grader, the unit costs realized for the entire finishing roadway item have averaged as follows:

Freeways in urban areas, 22 to 320 stations in total length

Labor $33 per station
Equipment $27 per station

Freeways in rural areas 1000 or more stations in total length

Labor $21 per station
Equipment $20 per station

City streets, with curbs and sidewalks, 50 stations in total length

Labor $39 per station
Equipment $14 per station

Small jobs in rural areas, 2 to 3 stations in total length

Labor $20 per station
Equipment $60 per station.

12

Drain Pipe

Culvert pipe and pipework relating to other surface drainage are normally installed by the general contractor. This work must usually be done in segments spread over the life of the job, and subcontracting is often not feasible, due to scheduling difficulties and excessive on and off costs.

Sewer lines and water piping are installed by subcontractors. When culvert pipes must be jacked under roadways, this work is also done by specialists, on a subcontract basis. In most cases the general contractor will furnish the jacking pit, which will be about 4 feet wider than the outside pipe diameter and about 4 feet longer than the pipe joint-length, and will have a vertical back face.

Pipe excavation and backfill are normally carried under separate job items, as discussed in Chapter 8. The unit costs herein for installing pipes include the costs for fine grading, but do not include the costs for excavation or for backfill.

12.1 Corrugated Steel Pipe and Pipe-Arches

The weights of typical riveted corrugated steel pipes are given in Tables 12.1 and 12.2. Table 12.1 covers pipes with 2⅔-inch pitch by ½-inch-deep corrugations; Table 12.2 covers pipes with 3-inch pitch by 1-inch-deep corrugations. Both tables are reproduced through the courtesy of Armco Inc., Metal Products Division. The standard joint length is 20 feet, and shorter pieces are prefabricated to order, as required. Typically, the suppliers take off the lengths required, and quote the material F.O.B. the job.

Table 12.1. Approximate Weights for Corrugated Steel Pipes

$2\frac{2}{3}'' \times \frac{1}{2}''$ Pipe[a]

Inside Diameter (inches)	Specified Thickness (inches)	Approximate Pounds per Linear Foot			
		Galva-nized	Full-Coated	Full-Coated and Invert Paved	Full-Coated and Full Paved
12	.064	10	12	15	
	.079	12	14	17	
15	.064	12	15	18	
	.079	15	18	21	
18	.064	15	19	22	
	.079	18	22	25	
	.109	24	28	31	
21	.064	17	21	26	
	.079	21	25	30	
	.109	29	33	38	
24	.064	19	24	30	45
	.079	24	29	35	50
	.109	33	38	44	60
30	.064	24	30	36	55
	.079	30	36	42	60
	.109	41	47	53	75
36	.064	29	36	44	65
	.079	36	43	51	75
	.109	49	56	64	90
	.138	62	69	77	100
42	.064	34	42	51	
	.079	42	50	59	85
	.109	57	65	74	105
	.138	72	80	89	115
48	.064	38	48	57	
	.079	48	58	67	95
	.109	65	75	84	120
	.138	82	92	101	130
	.168	100	110	119	155
54	.079	54	65	76	105
	.109	73	84	95	130
	.138	92	103	114	155
	.168	112	123	134	175
60	.079	60	71	85	
	.109	81	92	106	140
	.138	103	114	128	180
	.168	124	135	149	190

Table 12.1. (continued)

2⅔″ × ½″ Pipe[a]

Inside Diameter (inches)	Specified Thickness (inches)	Approximate Pounds per Linear Foot			
		Galva- nized	Full- Coated	Full-Coated and Invert Paved	Full-Coated and Full Paved
66	.079	65	77	93	
	.109	89	101	117	160
	.138	113	125	141	180
	.168	137	149	165	210
72	.109	98	112	129	170
	.138	123	137	154	210
	.168	149	163	180	240
78	.109	105	121	138	200
	.138	133	149	166	230
	.168	161	177	194	260
84	.138	144	161	179	240
	.168	173	190	208	270
90	.138	154	172	192	
	.168	186	204	224	
96	.168	198	217	239	

[a]Lock seam construction only; weights will vary with other fabrication methods.
Source: Armco Inc., Metal Products Division.

The weight per linear foot for a pipe-arch is equal to the weight of a pipe section having the same periphery. The diameter of the equivalent pipe section is approximately equal to the average of the two pipe-arch dimensions multiplied by 1.03.

Corrugated steel pipe is designated by wall thickness or by gage of the metal used. The relationship between wall thickness and gage is:

Wall thickness (inches)	Gage
0.064	16
0.079	14
0.109	12
0.138	10
0.168	8

Table 12.2. Approximate Weights for Corrugated Steel Pipes

$3'' \times 1''$ Pipe[a]

Inside Diameter (inches)	Specified Thickness (inches)	Approximate Pounds per Linear Foot			
		Galva-nized	Full-Coated	Full-Coated and Invert Paved	Full-Coated and Full Paved
36	.064	33	44	56	92
	.079	41	52	64	100
	.109	56	67	79	115
	.138	71	82	94	130
	.168	87	98	110	146
42	.064	39	52	66	107
	.079	47	60	74	116
	.109	65	78	92	134
	.138	83	96	110	152
	.168	100	113	127	169
48	.064	44	59	75	123
	.079	54	69	85	132
	.109	74	89	105	152
	.138	95	110	126	174
	.168	115	130	146	194
54	.064	50	66	84	138
	.079	61	77	95	149
	.109	83	100	118	171
	.138	106	123	140	194
	.168	129	146	163	217
60	.064	55	73	93	153
	.079	67	86	105	165
	.109	92	110	130	190
	.138	118	136	156	216
	.168	143	161	181	241
66	.064	60	80	102	168
	.079	74	94	116	181
	.109	101	121	143	208
	.138	129	149	171	236
	.168	157	177	199	264
72	.064	66	88	111	183
	.079	81	102	126	197
	.109	110	132	156	227
	.138	140	162	186	257
	.168	171	193	217	288
78	.064	71	95	121	198
	.079	87	111	137	214

Table 12.2. (continued)

$3'' \times 1''$ Pipe[a]

Inside Diameter (inches)	Specified Thickness (inches)	Approximate Pounds per Linear Foot			
		Galva-nized	Full-Coated	Full-Coated and Invert Paved	Full-Coated and Full Paved
78 (cont.)	.109	119	143	169	246
	.138	152	176	202	279
	.168	185	209	235	312
84	.064	77	102	130	213
	.079	94	119	147	230
	.109	128	154	182	264
	.138	164	189	217	300
	.168	199	224	253	335
90	.064	82	109	140	228
	.079	100	127	158	246
	.109	137	164	195	283
	.138	175	202	233	321
	.168	213	240	271	359
96	.064	87	116	149	242
	.079	107	136	169	262
	.109	147	176	209	302
	.138	188	217	250	343
	.168	228	257	290	383
102	.064	93	124	158	258
	.079	114	145	179	279
	.109	155	186	220	320
	.138	198	229	263	363
	.168	241	272	306	406
108	.079	120	153	188	295
	.109	165	198	233	340
	.138	211	244	279	386
	.168	256	289	324	431
114	.079	127	162	199	312
	.109	174	209	246	359
	.138	222	257	294	407
	.168	271	306	343	456
120	.109	183	220	259	378
	.138	234	271	310	429
	.168	284	321	360	479

[a]Lock seam construction only; weights will vary with other fabrication methods.
Source: Armco Inc., Metal Products Division.

Since the weight of a given joint will vary by more than 100 percent, depending on the gage, corrugations, and coatings, the costs to install the material are based on the weight per linear foot of pipe or of pipe-arch.

When corrugated steel pipes are installed, it is imperative that the outside circumferential laps be upstream, and longitudinal laps should not be located in the invert.

Average unit costs for installing plain galvanized corrugated steel pipe or pipe-arches are given by the following emperical formulas, which are based on cost data from many typical installations. These unit costs include the costs of unloading the material near the place of installation.

Equipment is not required to install pipes weighing up to about 30 pounds per linear foot, and

$$C_{LS} = 1.23 + 0.048W \qquad (12.1)$$

where

C_{LS} = labor cost in dollars per linear foot, indexed to K_L = 13.48, for pipes weighing from 8 to 30 pounds per linear foot.
W = the weight of the pipe or of the pipe-arch in pounds per linear foot.

The average unit costs to install the heavier pipes or pipe-arches are about

$$C_{LL} = 0.55 + 0.040W \qquad (12.2)$$

and

$$C_{RL} = 0.34 + 0.014W \qquad (12.3)$$

where

C_{LL} = labor cost, in dollars per linear foot, indexed to K_L = 13.48, for pipes weighing from 30 to 200 pounds per linear foot.
C_{RL} = the unit rental cost, corresponding to C_{LL}, indexed to K_{BH} = 25.
W = the weight of the pipe or of the pipe-arch in pounds per linear foot.

When bituminous coated pipe is specified, the unit weights for the coated pipe should be used in the preceding formulas, and the resulting unit costs should be increased by 25 percent.

Assuming the following conditions:

1. Plain galvanized corrugated steel pipe 36 inches in diameter by 12 gage, weight = 49 pounds per linear foot.
2. Hourly wage rate for a laborer = $15.00.
3. Hourly rental rate for a small rubber-tired backhoe = $29.00,

the estimated unit costs to install the material will be

$$C_{LL} = (0.55 + 0.040 \times 49) \frac{15.00}{13.48} = \$2.80$$

and

$$C_{LR} = (0.34 + 0.014 \times 49) \frac{29.00}{25} = \$1.20$$

Corrugated steel culverts normally have concrete headwalls at each end (minor structures—Section 4.17), or galvanized steel flared end sections.

Estimated Unit Costs to Install Plain Galvanized Steel Flared End Sections

Pipe diameter (inches)	Labor cost ($K_L = 13.48$)	Rental cost ($K_{BH} = 25$)	Unit of Measure
12–15	$ 40	—	Each
18	$ 50	—	Each
24	$ 60	—	Each
30	$ 75	—	Each
36	$ 60	$25	Each
42	$ 70	$25	Each
48	$ 80	$35	Each
60	$110	$50	Each

When bituminous coated end sections are specified, the preceding unit costs should be increased by about 25 percent.

Galvanized steel entrance tapers are normally used with 8- and 12-inch corrugated steel pipe down drains. Indexed to $K_L = 13.48$, the labor cost to install these is about $48 each for the 8-inch tapers and about $60 each for the 12-inch tapers. When bituminous coated tapers are specified, these costs should be increased by about 25 percent.

The preceding unit costs for installing flared end sections and entrance tapers include the costs for fine grading and all incidental excavation and backfill.

12.1.1 Salvaging Corrugated Steel Pipe

The unit labor cost to salvage corrugated steel pipe, whether plain galvanized or coated, will run about 2 to 2.5 times the unit labor cost for installing the plain galvanized material. The unit cost for equipment will be the same as the installation cost for plain galvanized pipe.

12.1.2   Corrugated Steel Multi-plate Structures

Multi-plate structures are used when large water areas are required. Prefabricated corrugated steel plates, together with the necessary bolts, are supplied by the manufacturers, and are assembled in place, on the job. Multi-plate pipes range from about 5 to 15 feet in diameter, multi-plate pipe-arches have spans ranging from about 6 to 17 feet, and multi-plate arch spans range from about 6 to 30 feet.

Multi-plate structures are erected by subcontractors specializing in this work, and the names of these subcontractors, for a given area, are obtained from the material suppliers.

12.2   Rubber Gasketed Reinforced Concrete Pipe

The weights per linear foot, the wall thicknesses, and the standard joint lengths for reinforced concrete pipes vary with the different manufacturers. Typical dimensions and weights for Class III rubber-gasketed reinforced concrete pipes are shown in Table 12.3.

Unit costs for installing rubber-gasketed reinforced concrete pipe, having 11-foot joint lengths, are shown in Table 12.4. It is assumed the pipe is unloaded by the supplier, at no additional cost, near the place of installation.

The unit costs shown in Table 12.4 assume average conditions, and that the trench depths are such that shoring is not required. When pipes are installed in shored trenches, the unit costs should be increased about 25

Table 12.3.   Rubber Gasketed Reinforced Concrete Pipes (Class III)

Inside diameter (inches)	Wall thickness (inches)	Weight (lbs. per linear foot)	Weight (tons per 11 foot joint)
12	2	105	0.58
18	2.25	170	0.94
24	2.50	255	1.40
30	2.75	350	1.93
36	3.125	475	2.61
42	3.75	675	3.71
48	4.125	850	4.68
54	4.5	1075	5.91
60	6	1375	7.56
66	6.5	1620	8.91
72	7	2000	11.0

Table 12.4. Unit Costs for Installing Rubber-Gasketed Reinforced Concrete Pipe (11-foot joint lengths)

Pipe Diameter (inches)	Unit Costs		Unit of Measure
	Labor ($K_L = 13.48$)	Rental[a] ($K_{Cr} = 75$)	
12	$1.30	$1.10	LF
18	$1.30	$1.30	LF
24	$1.30	$1.60	LF
30	$1.40	$1.90	LF
36	$1.50	$2.30	LF
42	$1.70	$2.80	LF
48	$2.10	$3.40	LF
54	$3.50	$5.20	LF
60	$4.90	$7.20	LF
66	$6.40	$9.20	LF
72	$7.80	$12.00	LF

[a]The costs for moving the crane on and off the job are additional.

percent. The unit costs include the costs for fine grading the trench, but do not include the costs for excavation or backfill, which are estimated as separate items.

Although the unit costs for rental are indexed to the hourly rental rate for a 35-ton truck crane, a smaller crane is satisfactory for the smaller pipe sizes, and a larger crane may be necessary for pipes over 5 feet in diameter. In all cases, the estimated unit cost for rental should be based on the hourly rental rate for a crane that can easily handle the pipe under the anticipated job conditions.

12.2.1 Hydrostatic Tests

Hydrostatic testing is normally required when rubber-gasketed reinforced concrete pipe is installed. After the pipe has been back-filled to a minimum depth of about 2 feet, the test will typically be conducted over a 24 hour period, with a 10 foot hydrostatic head on the highest point of the pipe. All obvious leaks must be stopped, and no leakage over the amount permitted by the specifications is allowed. The amount of leakage permitted is about 0.60 gallons per inch of inside diameter per 100 feet of pipe per hour.

For testing purposes, the ends of the pipe are normally sealed through the use of special end joints, which are furnished with concrete plugs in place. These test joints are supplied on a loan basis by the pipe manufacturer and, except for freight charges, no costs are involved. Nozzles are provided in the plugs for connecting the stand pipe and the supply hose, and for draining the line upon completion of the test.

The operations involved in making a hydrostatic test are:

1. The test joints must be installed and later removed.
2. The ends of the test joints must be strutted or backfilled to prevent the pipe joints from spreading under the hydrostatic head.
3. The line must be filled with water.
4. Any leaks which develop must be stopped.

Although testing costs are subject to considerable variation, typical costs when test joints are used are shown in Table 12.5.

In most cases, a water truck is used to fill the pipe; under average job conditions, the water truck averages about three loads per hour.

The amount of water required to fill a given pipe is

$$M = 0.00408 \, D^2$$

where

M = thousands of gallons per 100 linear feet of pipe.
D = pipe diameter in inches.

Under normal conditions, there is no additional labor cost to fill the section with water.

Table 12.5. Unit Costs for Hydrostatic Testing[a]

| Pipe Diameter (inches) | Unit Costs per Test | | |
	Labor ($K_L = 13.48$)	Rental ($K_{Cr} = 75$)	M-gallons per 100 LF
12	$130	$140	0.59
24	$160	$150	2.35
36	$200	$160	5.29
48	$270	$220	9.40
60	$380	$370	14.7
72	$520	$550	21.2

[a]The costs for furnishing the required water and for filling the section to be tested are additional.

It is sometimes necessary to seal the ends of a pipe line with timber bulkheads, when test joints are not available. Bulkheads for a 60-inch pipe will be installed and removed for a labor cost of about $420 per bulkhead, indexed to $K_{6+2} = 1000$. This cost assumes the length of the pipe line to be such that the bulkheads may be tied together with steel rods.

12.3   Tongue and Groove Reinforced Concrete Pipe

The weights per linear foot and the standard joint lengths for tongue and groove pipe vary with the different manufacturers. Typical weights for Class III tongue and groove pipes are shown in Table 12.6.

Unit costs for installing tongue and groove reinforced concrete pipe, having 8-foot joint lengths, are shown in Table 12.7. It is assumed the pipe is unloaded by the supplier, at no additional cost, near the place of installation.

The comments made following Table 12.4, Unit Costs for Installing Rubber-Gasketed Concrete Pipe, apply here also.

Tongue and groove pipe joints are sealed with a cement mortar band. The average quantities of cement required are shown in Table 12.7, and assume a mortar consisting of one part cement to two parts sand. About 0.12 tons of sand are required per sack of cement, but since the sizes of the mortar bands are not specified, the amounts of material required vary.

Table 12.6.   Tongue and Groove Reinforced Concrete Pipes (Class III)

Inside Diameter (inches)	Wall Thickness (inches)	Weight (lbs per linear foot)	Weight (tons per 8-foot joint)
12	2	100	0.40
18	2.25	160	0.64
24	2.50	230	0.92
30	2.75	315	1.26
36	3.125	425	1.70
42	3.875	625	2.50
48	4.25	775	3.10
54	4.625	950	3.80
60	5.25	1200	4.80
66	5.75	1450	5.80
72	6.25	1710	6.84
84	7	2590	10.4

Table 12.7.   Unit Costs for Installing Tongue and Groove Reinforced Concrete Pipe (8-foot joint lengths)

Pipe Diameter (inches)	Unit Costs		Cement (sacks per joint)	Unit of Measure
	Labor ($K_L = 13.48$)	Rental[a] ($K_{Cr} = 75$)		
12	$ 1.10	$ 0.80	0.05	LF
18	$ 1.60	$ 1.10	0.09	LF
24	$ 2.00	$ 1.40	0.13	LF
30	$ 2.60	$ 1.80	0.18	LF
36	$ 3.30	$ 2.30	0.23	LF
42	$ 4.20	$ 3.00	0.32	LF
48	$ 5.20	$ 3.90	0.41	LF
54	$ 6.30	$ 5.00	0.50	LF
60	$ 7.80	$ 6.50	0.59	LF
66	$ 9.50	$ 8.30	0.74	LF
72	$11.50	$10.60	0.89	LF
84	$17.20	$17.60	1.0	LF

[a]The costs for moving the crane on and off the job are additional.

The unit costs shown in Table 12.7 include the costs for the following operations:

1. Fine grading the trench.
2. Setting the pipe.
3. Installing and curing the mortar bands.

Although the equipment rental costs relate solely to the setting of the pipe, the labor cost breaks down about as follows:

1. Fine grading	22 percent
2. Setting	21 percent
3. Installing mortar bands	57 percent
	100 percent

12.4   Concrete Flared End Sections

The dimensions and weights of reinforced concrete flared end sections vary with different manufacturers. Typically, a small cast-in-place concrete cut-off wall 8 inches thick by 12 inches deep is poured at the end of

Table 12.8. Average Unit Costs to Install Reinforced Concrete Flared End Sections

Pipe Diameter (inches)	Unit Costs		Cut-off Wall (CY)	Unit of Measure
	Labor ($K_L = 13.48$)	Rental ($K_{Cr} = 75$)		
18	$ 30	$ 15		Each
24	$ 50	$ 25	0.2	Each
30	$ 65	$ 30		Each
36	$ 95	$ 45	0.3	Each
42	$140	$ 65		Each
48	$190	$ 90	0.4	Each
54	$240	$120		Each

the flared section. The unit costs shown in Table 12.8 include the costs of fine grading, unloading the end sections, incidental excavation and backfill, and the costs to form and pour the typical cut-off walls.

12.5 Cutting Reinforced Concrete Pipe

Concrete pipe is ordered to exact lengths, and field cutting is avoided when possible.

Table 12.9. Estimated Costs for Field Cutting Reinforced Concrete Pipe[a]

Pipe Diameter (inches)	Labor Cost per Cut ($K_L = 13.48$)	Time required (hours)
12	$ 12	0.5
24	$ 20	1
36	$ 40	2
48	$ 80	3
60	$190	4
72	$300	8

[a]The rental costs for the following equipment are additional:

1. Electric hand saw with carborundum blade.
2. Air compressor.
3. Paving breaker.

12.6   Shaped Bedding

Shaped bedding is sometimes specified for pipes installed under roadways. When required, shaped bedding is normally bid as a separate job item at a price per square foot of shaped surface.

The shaping is accomplished with a template conforming to the outside diameter of the pipe. The template is guided by two 2×4 headers which are set to the required line and grade.

Indexed to $K_{6+2} = 1000$, the headers will be set at a cost of about $0.75 per linear foot of header, and the template will cost about $7 per foot of width. Indexed to $K_L = 13.48$ and $K_{BH} = 25$, the unit costs for shaping and compacting will be about $0.32 per square foot for labor, and about $0.15 per square foot for equipment rental. In most cases, an allowance for template or header materials is not necessary.

13

Miscellaneous Costs

The unit costs for concrete removal and for a number of miscellaneous minor items and activities are covered in this chapter. The unit costs for the minor items are both interesting in themselves and useful for estimating unit costs for other work by comparison.

13.1 Concrete Removal

Subcontractors are always used when large quantities of concrete removal are required. Skilled equipment operators, expert planning and job supervision, and past experience on similar demolition work are essential. Concrete removal is inherently risky and a high degree of precision is not possible when estimating its cost. Because of this, and because of differing experience backgrounds, the prices quoted by experienced demolition contractors, for a given job, will sometimes vary by as much as 100 percent.

On unit price contracts, concrete removal is normally bid as a separate job item at a price per cubic yard, or as a lump sum. The price bid may include the costs of certain excavation and backfill, and normally includes the costs to load and dispose of the broken concrete; the costs for such work are covered in Chapters 7 and 8.

Production rates and unit costs for concrete removal on several jobs are given below. These rates relate only to the costs for breaking the concrete, ready for loading with a 2¼ to 3 CY loader. Unit costs or production rates should never be blindly used, and this is particularly true where demolition work is involved.

13.1.1   Paving Breaker Work

When paving breakers are used for breaking concrete, the moil points must be replaced with sharp ones at about 1-hour intervals. The sharpening must be done by experienced blacksmiths, since improperly annealed moil points break immediately.

Indexed to $K_L = 13.48$, assuming that the work surface is approximately horizontal and that an 80 pound paving breaker is used, reinforced concrete will be broken at a unit labor cost ranging from about $84 to $111 per cubic yard. Assuming one and one-half laborers per paving breaker, the corresponding production rates will range from about 0.24 to 0.18 cubic yards per hour.

If the paving breaker must be operated in a horizontal position, the preceding unit costs will increase from two to three times. In all cases, the rental costs for the compressor, hoses, and air tools; the cost for sharpening moil points; and the costs for staging or for supports for paving breakers working horizontally, are additional.

13.1.2   Wrecking Ball Work

Skilled workmen demolished two reinforced concrete T-girder bridges, using a crane equipped with a 5000 pound wrecking ball. 690 cubic yards of deck plus girders were broken at a rate of 8.3 cubic yards per hour. The hourly cost of the five-man crew was:

1 labor foreman	$13.48 + 1.35 = \$14.83$
1 laborer with cutting torch	$13.48 + 0.25 = \$13.73$
2 laborers with paving breakers	$2 \times 13.73 = \$27.46$
1 laborer helping	$\underline{\$13.48}$
	$\$69.50$

and the unit cost for labor (indexed to $K_L = 13.48$) was

$$\frac{69.50}{8.3} = \$8.40 \text{ per cubic yard}$$

Indexed to $K_{Cr} = 75$, and assuming the daily rental rate for the compressor and tools to be $50, the unit costs for the equipment were:

$$\text{Crane} \quad \frac{75}{8.3} = \$9.00 \text{ per cubic yard}$$

$$\text{Compressor} \quad \frac{6.25}{8.3} = \$0.75 \text{ per cubic yard}$$

780 cubic yards of bridge columns and footings were broken at a rate of 9.0 cubic yards per hour. The columns were approximately 3.5 feet square by 30 feet in height. A three-man crew was used; no paving breakers were required, although the compressor and tools were standing by.

Indexed to $K_L = 13.48$, $K_{Cr} = 75$, and to a daily rental rate of $50 for the standby compressor and tools, the unit costs were:

Labor $\qquad \dfrac{42.04}{9.0} = \$4.70$ per cubic yard

Crane $\qquad \dfrac{75}{9.0} = \$8.30$ per cubic yard

Compressor $\quad \dfrac{6.25}{9.0} = \$0.70$ per cubic yard

Indexed to the same labor and rental rates as above, a skilled five-man crew broke 380 cubic yards of 16-inch-thick flat slab bridge deck plus 95 cubic yards of abutments and wingwalls at unit costs of

Labor $\qquad$ $8.50 per cubic yard
Crane $\qquad$ $8.80 per cubic yard
Compressor $\quad$ $0.75 per cubic yard

70 reinforced concrete pile extensions, 16 inches in diameter by 7 feet long, were broken by the same crew at unit costs of

Labor $\qquad$ $145 per cubic yard, or $18 each
Crane $\qquad$ $126 per cubic yard, or $15 each
Compressor $\quad$ $ 13 per cubic yard, or $1.60 each

47 cubic yards of 6-inch, reinforced concrete deck slab, plus 35 cubic yards of abutments and wingwalls, were broken with a crane equipped with a 5000-pound wrecking ball at a rate of 5.8 cubic yards per hour. The crane operator was average, and one laborer was used as a burner. No paving breakers were required. Indexed to $K_L = 13.48$, and $K_{Cr} = 75$, the unit costs were:

Labor $\quad \dfrac{13.48 + 0.25}{5.8} = \$ 2.40$ per cubic yard

Crane $\qquad \dfrac{75}{5.8} = \$13.00$ per cubic yard

Indexed to K_L = 13.48, K_{Cr} = 75, and a daily rental rate for a compressor and paving breaker of $50, 44 reinforced concrete headwalls averaging 1.38 cubic yards each were cut loose and broken for unit costs of

Labor	$37 each, or $27 per cubic yard
Crane	$56 each, or $41 per cubic yard
Compressor	$11 each, or $ 8 per cubic yard

13.1.3   Hydraulic-Impulse Breakers

These units weigh about 1000 pounds and are mounted on the boom of a small rubber-tired loader-backhoe. They operate from the tractor's hydraulic system, and no compressor is required.

42 cubic yards of reinforced concrete, consisting of the outer 3 feet of a 12-inch-thick bridge deck overhang, and a 2-foot walkway 8 inches thick, on a deck slab which was to remain, were removed. One and one-half laborers, one paving breaker and one hoe-mounted impulse-breaker were used.

Indexed to K_L = 13.48, to an hourly rental rate for the impulse-breaker of $40, and to a daily rental rate for the compressor and paving breaker of $50, the unit costs were:

Labor	$51 per cubic yard
Impulse breaker	$67 per cubic yard
Compressor	$ 7 per cubic yard

In all cases, when portions of a reinforced concrete structure are to be removed, extra care and time are required to avoid damaging the remaining structure.

13.1.4   Concrete Sawing

Saw-cutting is frequently required when portions of a concrete structure must be removed. This work is usually done by subcontractors using self-propelled concrete cutting saws equipped with diamond blades, and the same equipment is used to saw-cut weakened-plane joints in new concrete pavement.

Saw-cuts in old unreinforced concrete slabs are made at about the following speeds.

Depth of cut	Linear feet per hour
1½″	55
2″	50
3″	40
4″	25

Actual saw-cutting speeds vary with the quality of the concrete and with the amount of reinforcing steel encountered. With #4 or #5 bars at 12-inch centers, the cutting speeds will range from about 35 to 50 percent of those shown above. When asphalt concrete or fresh portland cement concrete is cut, the speeds will be about 6 times those shown above.

Vertical and horizontal cuts in reinforced concrete walls are usually short, and require more time. The self-propelled saw used is lighter in weight, and operates from a temporary rail which must be bolted to the wall. Door openings, 3 feet by 7 feet, will be cut in 4-inch reinforced concrete walls in about 4 hours, and about 6 hours are required for such openings in walls 7 inches thick.

13.1.5   Pavement Removal

A 185-horsepower tractor ripped 8½-inch-thick asphalt concrete pavement at an effective rate of about 550 square yards per hour. Cracked portland cement concrete pavement, 8 inches thick, was ripped at an effective rate of about 340 square yards per hour. In both cases, the material was loaded with a track-type loader having a 2½-cubic-yard bucket.

Portland cement concrete pavement 8 inches thick was broken by a crane equipped with a 5000-pound wrecking ball at an effective rate of about 139 square yards per hour. Using a self-propelled, 1500-pound hydraulic hammer, the same pavement was broken at an effective rate of about 220 square yards per hour. In both cases the material was loaded with a track-type loader having a 2½-cubic-yard bucket.

A crane equipped with a wrecking ball broke concrete curb, gutter, and sidewalk at an effective rate of about 77 cubic yards per hour.

13.2   Minor Items and Costs

The costs for the following items are, in most cases, small in comparison with the total job cost. The data presented represents average costs when incidental amounts of such work are done by the general contractor's own forces.

13.2.1 Markers and Reflectors

A typical highway marker consists of a 10 gauge galvanized post 3 inches wide by 6 feet long, having a single vertical corrugation, and weighing about 15 pounds. The posts are driven to a penetration of 2 feet at specified locations. An enameled plate, 8 by 24 inches, weighing about 7 pounds, is attached to the post with eight ¼-inch bolts.

Indexed to K_L = 13.48, the labor cost to install markers, complete, is about $4.65, and no equipment rental is involved.

13.2.2 Right-of-Way Monuments

A typical right-of-way monument is a portland cement concrete post 6 inches square by 3 feet 6 inches long, weighing about 125 pounds. The posts are set 3 feet deep at specified locations.

Indexed to K_L = 13.48, the labor cost to set each prefabricated monument is about $19.

13.2.3 Survey Monuments

Survey monuments are of two types. A type 1 monument is a concrete post about 15 inches square by 24 inches long, weighs about 450 pounds, and is set flush with the ground. A type 2 monument is about 15 inches square by 18 inches long, weighs about 370 pounds, and is set 28 inches deep. A short piece of 6-inch-diameter pipe with a cover is set on top of the post.

Indexed to K_L = 13.48, the labor cost to set either prefabricated monument will be about $20. Type 1 monuments, cast in place, will cost about $30 each for labor.

13.2.4 Metal Beam Guard Rail

Metal beam guard rail consists of a corrugated metal rail bolted to wood posts. The galvanized rail panels are about 12 inches wide by 13 feet six inches long, have two 3¼-inch corrugations, and weigh about 108 pounds. The treated posts are 8×8 rough by 5 feet 4 inches long, and are embedded 3 feet deep at 6 foot 3 inch or 12 foot 6 inch centers. An 8×8 filler block 14 inches long is used at each post, and the rail is attached with one ⅝-inch bolt per post. The rail panels are lap spliced at the posts and eight ⅝-inch bolts are used for each splice.

Indexed to K_{2+3} = 585, where K_{2+3} = the total cost for two carpenters plus three laborers working for 8 hours, the unit labor costs for installing metal beam guard railing are about as follows:

>6'-3" post spacing: $4.16 per linear foot
>(140 linear feet per day)
>12'-6" post spacing: $3.92 per linear foot
>(150 linear feet per day).

In both cases, a hand operated, power-driven post hole auger is required; when large quantities are involved, a truck mounted auger may be used.

13.2.5 Right-of-Way Fence

Fencing is always installed by subcontractors when the quantities are large. Two laborers installed 5-strand barbed wire fence on corrugated metal posts spaced 12 feet on centers, at a rate of 480 linear feet per day. Indexed to K_L = 13.48, the unit labor cost was $0.45 per linear foot. Using wood posts, a similar fence was installed for a unit labor cost of $0.95 per linear foot, and a hand-operated, power-driven post hole auger was required.

13.2.6 Raised Traffic Bars

Raised traffic bars are prefabricated of portland cement concrete, are 8 inches wide by 1½ to 2 inches high, and weigh 9 to 14 pounds per linear foot. They are installed on a thick trowel coat of asphalt adhesive, and are given one coat of traffic paint on completion.

Indexed to K_L = 13.48, small quantities of raised traffic bars were installed, complete, at a unit labor cost of $1.10 per linear foot.

13.2.7 Barricades

Indexed to K_{6+2} = 1000, portable timber barricades 4 feet wide by 5 feet high were made from 2-inch material at a unit labor cost of $260 per MFBM. Barricades 12 feet wide by 5 feet high were made from 2- to 4-inch material at a unit labor cost of $270 per MFBM. In both cases, the costs include the application of one coat of paint.

On unit price contracts, the price bid for barricades normally includes the costs for furnishing, placing, repairing, or replacing damaged units. In general, the longer the barricades are in use, the more damage they sustain, and on typical freeway work, more than two barricades must be furnished, over the life of the job, for each one covered by the job item.

13.2.8 Roadside Signs (Wood Posts)

The sign panels for roadside signs are normally furnished, F.O.B. job, by the owner. The price bid for roadside signs includes the furnishing of all wood posts, braces, and hardware, and the costs for installing the posts and sign panels, complete, in place. The unit of measure for the item is thousands of board feet.

Indexed to $K_{6+2} = 1000$, typical roadside signs having 4×4 to 6×8 wood posts were installed at an average unit cost for labor of $1070 per MFBM, not including the cost of painting. A hand-operated, power-driven post hole auger was used. Indexed to $K_L = 13.48$, the unit labor cost for painting the posts was about $330 per MFBM for three coats.

13.2.9 Roadside Signs (Steel Posts)

This item is similar to the one above, except that the unit of measure is the pounds of steel in the posts. In addition, the steel posts are normally backfilled with 4 sack concrete.

Indexed to $K_{6+2} = 1000$, and $K_{Cr} = 50$, roadside signs on steel posts were installed at the following unit costs:

Labor $96 per post or $0.16 per pound
Crane $33 per post or $0.055 per pound

A truck-mounted post hole auger was used; it averaged four holes per hour.

13.2.10 Remove Roadside Signs (Wood Posts)

Indexed to $K_L = 13.48$, roadside signs with wood posts were removed and stockpiled at a unit cost for labor of $8.00 per post. No equipment was required.

13.2.11 Sacked Concrete Slope Protection

Sacks made of 10-ounce burlap, measuring about 19½ inches wide by 36 inches long, are filled with 1 cubic foot of 4 sack concrete, and placed on the slope as specified.

This is heavy work, and it is imperative that the concrete be delivered close to the location where the sacked concrete will be placed.

Using a portable hopper, equipped with a job-made 1-cubic-foot metering box, 10 laborers placed several hundred cubic yards of sacked concrete at an average rate of 3.1 cubic yards per hour. Two three-man crews were used to carry and place the sacks, and the average carry distance was about 15 feet, down a 2:1 slope. One laborer was used grading and filling at the point of depositing, and three laborers filled the sacks at the hopper and handed them to the carrying crews.

Indexed to $K_L = 13.48$, the unit labor cost for this work was $44 per cubic yard.

When small quantities are involved, and when the concrete can be chuted directly to the point of deposit from the truck, a four-man crew will do the work at a slower rate for the same unit cost. These unit costs do not include the costs for excavation, which is normally paid for under a separate job item.

13.2.12 Sand Bags

Three laborers can fill and place about 40 bags per hour if the bags can be deposited close to the point of filling. The bags contain about 0.9-cubic feet of sand, and will cover about 1.8 SF per bag on the level, and about 1.4 SF per bag on a 1½:1 slope.

13.2.13 Grouting Broken Concrete Riprap

Indexed to $K_L = 13.48$, broken concrete riprap was grouted with concrete, which was chuted to the required spot, at a unit labor cost of $12 per cubic yard. This unit cost includes the costs of spading the grout into the spaces between the pieces of broken concrete, finishing the surface so the faces of the broken pieces are exposed, and curing the grout.

13.2.14 Rock Slope Protection

Indexed to $K_L = 13.48$, and $K_{Cr} = 75$, 270 cubic yards of ¼-ton rock slope protection was placed at unit costs of:

Labor $1.40 per cubic yard
Rental $4.20 per cubic yard

The material was end dumped over a bank and placed with a ¾-cubic-yard clamshell bucket.

Indexed to the same rates, 270 cubic yards of excavation, for the rock, was done at unit costs of

Labor $0.65 per cubic yard
Rental $1.60 per cubic yard

400 tons of rock backing, consisting of rocks weighing from 25 to 50 pounds each, were placed with a 1-cubic-yard clamshell bucket at an effective rate of 40 tons per hour. 700 tons of 1-ton rock slope protection was placed over the backing, with the same equipment, at an effective rate of 43 tons per hour. In both cases, the material was placed on a relatively inaccessible slope behind a concrete pile bent which was about 15 feet in height. Working on banks approximately 8 feet in height, both the backing and the 1-ton rock were placed with a 2¼ cubic yard wheel loader at an effective rate of 60 tons per hour.

13.2.15   Dry Pack Window Mullions

Window mullions, requiring approximately 2 cubic inches of dry pack per mullion, were dry packed at a unit cost of $2.86 each, indexed to $K_F = 16.10$. This is equivalent to an effective rate of

$$\frac{16.10}{2.86} = 5.6 \text{ mullions per hour}$$

and this rate is typical for many repetitive operations of this type, where the workman must go to a location, do a minimum amount of work, clean up, and move to the next nearby location.

13.2.16   Welding

Pieces of iron pipe, 2 inches in diameter by 10 inches long, were welded to the top flanges of plate girders for use as screed supports. The deck forms were in place, and access was easy. The pipes were set 4 feet on center, and four spot welds were made for each pipe. A good welder attached the pipes at a rate of 36 pipes per hour.

Angle iron clips were welded to the top flanges of the outside girders at 4-foot centers. Eight inches of ¼-inch fillet weld was used for each bracket, and a good welder did the work at a rate of 7 brackets per hour.

13.2.17   Remove Wooden Rail Posts

8×8 bridge rail posts, 5 feet long, were removed and stockpiled for salvage. Each post had four rusted bolts which were burned and driven out with drift pins. Two laborers did the work at a rate of 2.8 posts per hour.

13.2.18 Wood Box Cofferdam

150 linear feet of wood box cofferdam, filled with earth, was used to dry up one side of an unlined canal. The water was 4 feet deep.

The box section was 4 feet wide by 5 feet high, and contained approximately 110 cubic yards of dirt. 2×12 planks were used for the box sides. 4×6 vertical wales together with two 4×4 struts were used at 4-foot centers, and the wales were tied together with twisted No. 9 wire. The open-ended box sections averaged 14 feet in length, and the entire cofferdam required 4.5 MFBM of material, which is equivalent to 3 board feet per square foot of box wall.

Indexed to $K_{6+2} = 1000$ and $K_{Cr} = 75$, the unit costs were:

Make box sections

Labor $250 per MFBM

Set box sections and fill with dirt

Labor $430 per MFBM
Rental $230 per MFBM

Remove box sections and dirt

Labor $310 per MFBM
Rental $340 per MFBM

The dirt fill was placed and partially removed with a ¾ CY clamshell bucket. The final clean up was made with a ¾ CY dragline bucket.

13.2.19 Drilling Dowel Holes in Concrete

The cost to drill holes in existing concrete structures depends on the amount of reinforcing steel encountered. Using a light jackhammer, holes 1½ inches in diameter were drilled vertically in unreinforced concrete at a rate of 120 inches per hour. Similar holes were drilled horizontally in lightly reinforced concrete at a rate of 48 inches per hour, and in heavily reinforced concrete at a rate of 26 inches per hour.

One laborer will dry-pack dowels in 24-inch-deep drilled holes at a rate of about 6 dowels per hour.

Index

397